Bioaccumulation in Aquatic Systems

Edited by R. Nagel and R. Loskill

VCH

Distribution:

VCH, P. O. Box 10 11 61, D-6940 Weinheim (Federal Republic of Germany)

Switzerland: VCH, P. O. Box, CH-4020 Basel (Switzerland)

United Kingdom and Ireland: VCH (UK) Ltd., 8 Wellington Court, Cambridge CB1 1HZ (England)

USA and Canada: VCH, Suite 909, 220 East 23rd Street, New York, NY 10010-4606 (USA)

ISBN 3-527-28395-1 (VCH, Weinheim) ISBN 1-56081-201-X (VCH, New York)

Bioaccumulation in Aquatic Systems

Contributions to the Assessment

Proceedings of an International Workshop, Berlin 1990

Edited by R. Nagel and R. Loskill

VCH Weinheim · New York · Basel · Cambridge

Editors:
Dr. Roland Nagel
Dr. Renate Loskill
Institute of Zoology
University of Mainz
P.O. Box 3980
D-6500 Mainz
Federal Republic of Germany

This book was carefully produced. Nevertheless, authors, editors and publisher do not warrant the information contained therein to be free of errors. Readers are advised to keep in mind that statements, data, illustrations, procedural details or other items may inadvertently be inaccurate.

1st edition 1991

Published jointly by
VCH Verlagsgesellschaft mbH, Weinheim (Federal Republic of Germany)
VCH Publishers Inc., New York, NY (USA)

Editorial Director: Dr. Hans-Joachim Kraus
Production Manager: Dipl.-Wirt.-Ing. (FH) Bernd Riedel

Library of Congress Card No. applied for

British Library Cataloguing-in-Publication Data
Bioaccumulation in aquatic systems: Contributions
to the assessment.
I. Nagel, N. II. Loskill, R.
574.92
ISBN 3-527-28395-1

Die Deutsche Bibliothek – CIP-Einheitsaufnahme
Bioaccumulation in aquatic systems ; contribitions to the assessment ; proceedings of an international workshop, Berlin 1990 / ed. by R. Nagel and R. Loskill – 1. ed. – Weinheim ; New York ; Basel ; Cambridge : VCH, 1991
ISBN 3-527-28395-1
NE: Nagel, Roland [Hrsg.]

Printed on acid-free and low-chlorine paper.

Printing: betz-druck gmbh, D-6100 Darmstadt 12
Bookbinding: Wilh. Osswald + Co., D-6730 Neustadt/Weinstraße
Printed in the Federal Republic of Germany

Preface

The bioaccumulation of chemicals is an important element in the assessment of environmental hazards. In order to enable a comprehensive evaluation of bioaccumulation, an international workshop and a complementary literature study were performed. The workshop "Contributions to the Assessment of Bioaccumulation of Organic Chemicals in Aquatic Systems" took place in Berlin on December 6.-7., 1990 and was financially supported by the Umweltbundesamt. The book is a proceeding of this workshop.

The development of the book has been supported by Dr. H. J. Kraus, VCH Verlagsgesellschaft, Weinheim (FRG), and we are thankful for his help. We also thank the authors for submitting their manuscripts.

R. Nagel
R. Loskill

Mainz, August 1991

Contents

List of Contributors

R. Anliker
Ecological and Toxicological Association of the Dystuffs
Manufactoring Industry (ETAD)
P.O.Box
CH-4005 Basel, Switzerland

B. Beek
Federal Environmental Agency
Bismarckplatz 1
1000 Berlin 33, FRG

S. Böhling
Federal Environmental Agency
Bismarckplatz 1
1000 Berlin 33, FRG

W. Butte
Fachbereich Chemie, Universität Oldenburg
D-2900 Oldenburg, FRG

N. Caspers
Bayer AG, WV Umweltschutz
D-5090 Leverkusen, Bayerwerk, FRG

D.W. Connell
Division of Australian Environmental Studies
Griffith University
Nathan, Queensland 4111, Australia

W. Ernst
Alfred Wegener Institute for Polar and Marine Research
Bremerhaven and University of Bremen
D-2850 Bremerhaven, FRG

P.D. Hansen
Institute for Ecology
Technical University of Berlin
D-1000 Berlin, FRG

P. Kristensen
The Water Quality Institute
Agern Alle 11
DK-2970 Horsholm, Denmark

R. Loskill
Universität Mainz
Institut für Zoologie
D-6500 Mainz, FRG

J.M. McKim
U.S. Environmental Protection Agency
Environmental Research Laboratory-Duluth
6201 Congdon Boulevard
Duluth, MN 55804, USA

P. Moser
Physical Chemistry, Central Research Services FD 3.1
CIBA-GEIGY A.G.
CH-4002 Basel, Switzerland

R. Nagel
Institut für Zoologie
Universität Mainz
D-6500 Mainz, FRG

M. Nendza
Fraunhofer-Institut für Umweltchemie und Ökotoxikologie
D-5948 Schmallenberg, FRG

A. Opperhuizen
Environmental Toxicology Section
Research Institute of Toxicology
University of Utrecht
P.O.Box 80.176
NL-3508 TD Utrecht, The Netherlands

H. Rosenthal
Institute for Marine Research and University of Kiel
D-2300 Kiel, FRG

P.K. Schmieder
U.S. Environmental Protection Agency
Environmental Research Laboratory-Duluth
6201 Congdon Boulevard
Duluth, MN 55804, USA

G. Schüürmann
Bayer AG, WV Umweltschutz
D-5090 Leverkusen, Bayerwerk, FRG

D.T.H.M. Sijm
Environmental Toxicology Section
Research Institute of Toxicology
University of Utrecht
P.O.Box 80.176
NL-3508 TD Utrecht, The Netherlands

H. Tyle
The National Agency of Environmental Protection
Strandgade 29
DK-1401 Kobenhavn, Denmark

S. Weigelt
Köhlerstraße 6
D-2000 Hamburg-Osdorf, FRG

Welcome Address and Introduction

Bernd Beek

Ladies and Gentlemen,

it is a great pleasure for me to welcome you to the Federal Environmental Agency in Berlin, and I am especially happy that so many well-known experts in the field from different countries are willing to contribute to this workshop on "The Assessment of Bioaccumulation of Organic Chemicals in Aquatic Systems".

It is generally accepted that Bioaccumulation of chemicals is an important factor in the assessment of environmental hazards, and it has been used in fact already for quite a long time as a trigger factor for decisions of administrative relevance.

Among others, the main areas of legislation and administration in the Federal Republic of Germany, in which Bioaccumulation is of importance either as such or as an additional factor are the following:

- The Chemicals Act
- The Plant Protection Act
- The Detergents and Cleansing Act
- The Federal Immission Control Act
- The Technical Instruction on Air Pollution Control
- The Federal Water Act

However, various deficiencies in our knowledge as to what Bioaccumulation really implicates in consequence, have become evident during the recent years, and it is our goal that this workshop will bring us at least a few steps further towards a solution of some problems related to the assessment of Bioaccumulation.

According to the implementation of the Chemicals Act, Bioaccumulation is assessed in a "three-tier-system" (Figure 1.1), related to the production volume of a particular chemical. In the so-called Base-set, ranging from 1 to 100 tons of production and distribution per year, the potential for Bioaccumulation is estimated from physico-chemical properties of the substance, in particular from the partition coefficient n-octanol/water, which is expressed as the log P_{ow} value:

- No Bioaccumulation is assumed below a log P_{ow} of 2.7.
- Bioaccumulation is assumed for the substances with a log P_{ow} between 2.7 and 6.0. Within this range a linear correlation with Bioconcentration Factors (BCF) may be obtained.
- Bioaccumulation is assumed for substances with a log P_{ow} higher than 6, but a molecular weight (MW) below 500*. A linear correlation with BCF, however, may no longer be obtained.

The latter two conditions may indicate, in combination with other factors, the necessity for testing the Bioaccumulation in fish.

- No considerable Bioaccumulation is assumed for substances with a log P_{ow} higher than 6 and a MW over 500*, in spite of a high lipophilicity. This is mainly because of the big size and other structural features (configuration) of the molecule, which inhibit or exclude penetration through biological membranes.

At Step I, ranging from 100 to 1000 tons of production and distribution per year, in any case, that means independent from the log P_{ow} values obtained at the Base-set level, a Bioaccumulation study, preferably in fish, is claimed. The philosophy behind this is the idea that log P_{ow} values in general may only serve as indicators, and can not replace a Bioaccumulation study as such.

At even higher production levels, that means, from 1000 tons per year onwards, at the so-called Step II, further Bioaccumulation studies may be necessary. On a case by case basis, the testing programme is designed in a "dialogue-phase" in interaction with the notifier.

In the Assessment of Bioaccumulation according to the implementation of the Plant Protection Act (Figure 1.2), the log P_{ow} value is solely used as a trigger factor: After determination of the log P_{ow} value, a Bioaccumulation study in fish is immediately claimed when the log P_{ow} is equal to or higher than 3. In case of a log P_{ow} value lower than 3, no Bioaccumulation study is performed.

---

* According to very recent developments in the field, we consider to propose a MW below 600.

BASE SET (≥ 1t/y)	**PARTITION COEFFICIENT** **n-octanol/water (log Pow)**

log Pow < 2.7	log Pow 2.7 - 6.0	log Pow > 6
No indication for Bioaccumulation	Indication for Bioaccumulation. Linear Correlation with BCF	Indication for Bioaccumulation. No longer linear Correlation with BCF

May indicate Bioaccumulation Test before Step I (in combination with other factors)

log Pow > 6
and
MW > 500
No considerable Bioaccumulation assumed

STEP I (≥ 100 t/y)	**BIOACCUMULATION TEST** (Mostly in Fish)
STEP II (≥ 1000 t/y)	**FURTHER BIOACCUMULATION STUDIES** ("Dialogue-phase")

Figure 1.1 The Assessment of Bioaccumulation - Chemicals Act
(A tiered system according to production volume)

PARTITION COEFFICIENT
n-OCTANOL/WATER (log Pow)

log Pow < 3	**log Pow > 3**
No Bioaccumulation Study	Immediately Bio-accumulation in Fish

Figure 1.2 The Assessment of Bioaccumulation - Plant Protection Act

Depending on the specific results of a Bioaccumulation study, further more sophisticated investigations related to, for example, tissue specific accumulation, metabolism, clearance, etc., may become necessary.

It must be pointed out that deviations from the two schemes described are of course possible, if they are scientifically justified.

From our practical experience in implementing the Chemicals Act and the Plant Protection Act, the assessment of Bioaccumulation is important not only in terms of quality, but also in terms of quantity. In Figure 1.3 the distribution of log P_{ow} values is shown, obtained from notifications according th the Chemicals Act (Base-set). Whereas in 54% of the cases no indication for Bioaccumulation has been obtained, in as much as the remaining 46% Bioaccumulation of the substances had to be assumed. From these, 34% fall into the class of linear correlation with BCF (log P_{ow}: 2.7 - 6.0), and 12% into the class of a still high Bioaccumulation potential (log P_{ow} higher than 6), but with no further linear correlation with BCF (The discrimination of substances with a log P_{ow} higher than 6 and a MW of higher than 500 as a different class, where no considerable Bioaccumulation is assumed, has been established only during the very recent years. These up to now not very frequent cases are included in class log P_{ow} higher than 6).

In Figure 1.4, the results are shown of an evaluation of 139 active ingredients gained from the clearance procedure of the Plant Protection Act. To enable a more direct comparison with the results obtained from the Chemicals Act-notifications, we chose a log P_{ow} value of 2.7 rather than of 3 for discrimination. As shown in Figure 1.4, 54.7% of active ingredients in Plant Protection Agents evaluated have a potential to bioaccumulate.

These large quantities of substances released into the environment with a potential for Bioaccumulation - either new chemicals or active ingredients in Plant Protection Agents - additionally point to the urgent need for solving problems related to the assessment of Bioaccumulation.

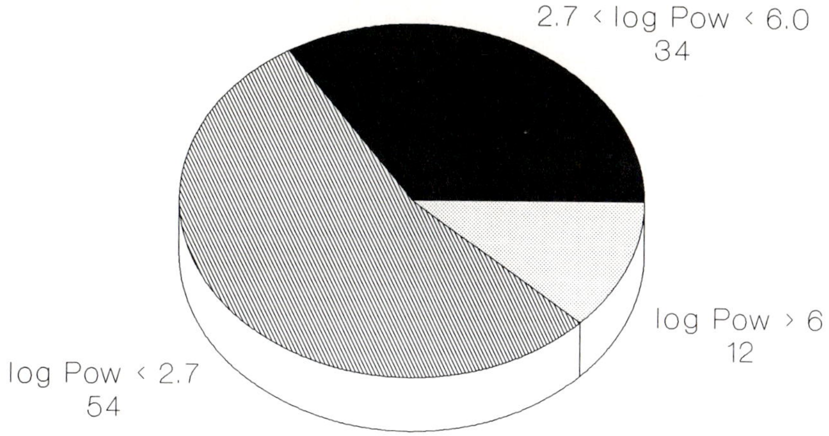

Figure 1.3 Distribution of log P_{ow} values (%) - Chemicals Act

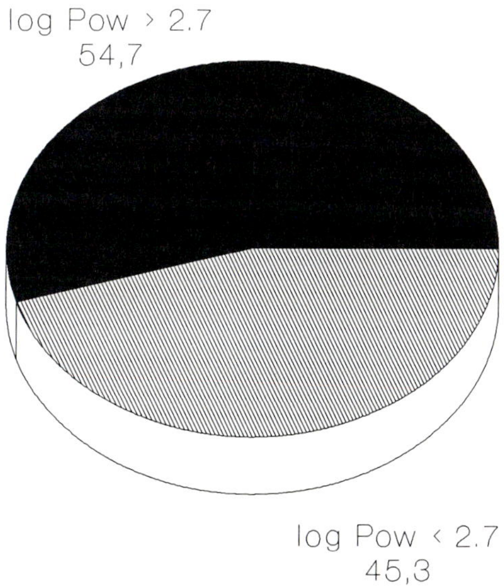

Figure 1.4 Distribution of log P_{ow} values (%) - Plant Protection Act

Purpose of the Workshop

Stella Böhling and Renate Loskill

2.1 Introduction

Compounds exhibiting a potential to accumulate are considered as critical, even if they are designated as non hazardous according to toxicity tests (cf. Hose et al. 1986, Borgmann et al. 1990). Toxicity may be acute and kill the organisms rather quickly, or chronic having gradual effects upon activity, feeding, reproduction and general physiology. The first is much more obvious - mortality of large numbers of fish soon attracts attention - the latter is much harder to detect; often it is not possible to quantitate or define a particular risk with certainty. The ecotoxicological effects assessment is rendered more difficult by the process of bioaccumulation (Figure 2.1). Concentrations of compounds, that according to bioassay criteria for acute or even chronic exposure appear to be safe, can accumulate to levels that are harmful to the organisms themselves or to the consumers of such organisms. When toxic thresholds are high, chronic effects from residue-forming chemicals may not be noticed until significant amounts have been released into the environment. Additionally, it must carefully be considered, whether the transfer from organism to organism through the food web to higher trophic levels will affect predators which might be more sensitive. The latter phenomenon is especially significant when considering the residues of some pesticides obtained through such predator-prey relationships, or PCB accumulation with resultant adverse effects (eg. Hoffman et al. 1990, cf. Broekhuizen and De Ruiter-Dijkman 1988).

Moreover, hazardous effects of chemicals stored in an organism may be induced by environmental as well as physiological changes such as hunger, when these compounds are liberated from their depot compartments, thus possibly exceeding toxic threshold concentrations (Frische et al. 1979, cf. Boon and Duinker 1985).

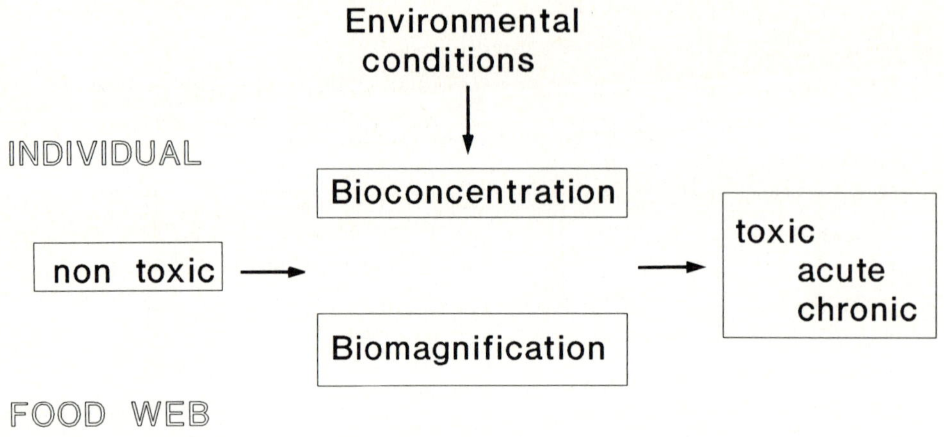

Figure 2.1 Environmental Relevance of Bioaccumulation

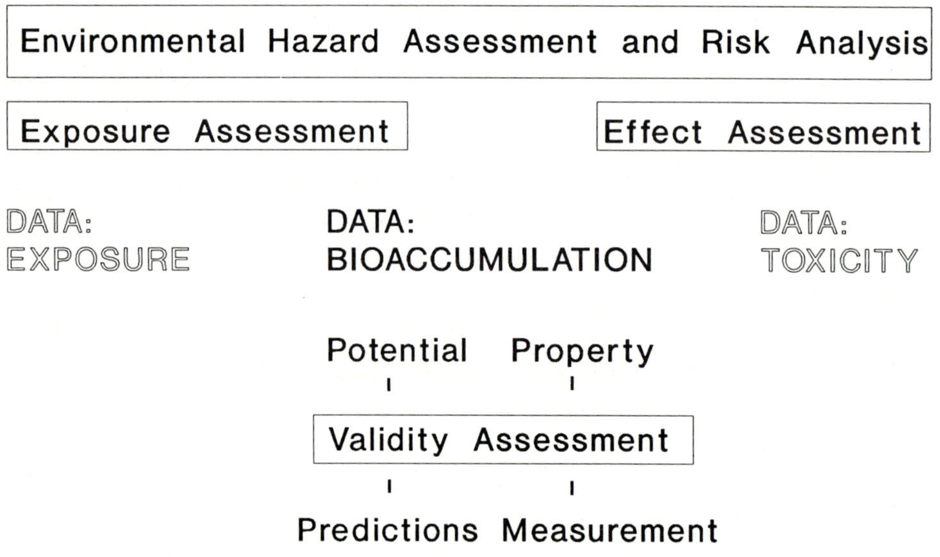

Figure 2.2 Bioaccumulation - an Essential Step in Environmental Hazard and Risk Assessment

In German environmental legislation the assessment of bioaccumulation data is considered an essential step in environmental risk and hazard assessment (Figure 2.2). This applies to

* the ecotoxicological assessment of industrial chemicals (new substances as well as existing substances), with respect to their impact on aquatic systems within the framework of the Chemicals Act as of 16.9.80.
* the assessment of ecological effects of plant protection agents, especially to non-target organisms, within the framework of the Plant Protection Act as of 15.9.86.
* the assessment of the impact of ingredients of washing and cleansing products on surface water (but also on sewage plants and on drinking water supply), within the framework of the Detergents and Cleansing Agents Act as of 5.3.87.

During the course of this workshop, the considerations will be limited to organic compounds, essentially to those substances above, for which an ecological effects assessment is prescribed by German law. Nevertheless, there is no doubt that bioaccumulation of heavy metals and organometallic compounds are of at least equal importance, as recently reported by Hawker (1990). Furthermore, only aquatic systems will be in the centre of interest, as there are relevant differences when compared to the terrestrial environment, the major difference being the situation of exposure (for review see Walker 1990). Terrestrial organisms are exposed to air, water, soil or food or combinations of these media. As opposed to direct uptake from the ambient medium for aquatic organisms, indirect bioaccumulation by oral uptake is the dominant mechanism effective in the terrestrial environment, the dermal route might contribute to this uptake in animals inhabiting soil. Significant uptake by the respiratory route will however rather be the exception. Studies on the bioaccumulation behaviour of terrestrial organisms are an important field of recent research, and it also holds true for terrestrial ecosystems, that comprehensive concepts for the assessment of bioaccumulation have not yet been developed. However, this cannot be achieved within the limited scope of this workshop and therefore in the following can only be subject to minor consideration.

2.2 Data Basis for the Assessment of Bioaccumulation

Within the framework of the tasks performed by the UBA especially in implementing the Chemicals Act and Plant Protection Act, it receives a large number of data on bioaccumulation. In general these are on the basis of the bioconcentration factor (BCF), a proportionality constant relating the concentration of a chemical in an organism to its concentration in the ambient water at steady state (Kenaga 1972). Bioconcentration is usually related to weight of whole body or specified tissues thereof, for example muscle or liver. So far, only few studies relate the BCF to the lipid weight.

* On the one hand, BCF values result from experimental procedures, i.e. internationally standardized tests (using static and semi-static systems as well as flow-through systems) or non standardized tests, including various different model ecosystems, field studies, monitoring programmes and case studies. At present, the five OECD-Guidelines are the most frequent in use in bioaccumulation testing; these have been used practically without any revision since their adoption in 1981 (OECD 1981). However, there are current efforts to update the OECD 305 E on the one hand as well as to consolidate the five guidelines into one by harmonizing basic test conditions.

Results from non-standardized tests raise several questions which must carefully be considered when attempting to assess such data. Besides usually lacking reproducibility, measured BCFs are in these cases not necessarily steady state values, thus not representing maximum bioconcentration potential and consequently making any comparison of results practically impossible. A further issue of interest in this context is which significance can be ascribed to field observations (incl. massive fish kills) and monitoring results vis-à-vis laboratory results?

* On the other hand, BCFs are predicted on the basis of QSARs, the most important physico-chemical property for estimating the bioconcentration potential of a chemical currently being its n-octanol/water partition coefficient, a measure of chemical hydrophobicity. Between the log P_{ow} and the log BCF numerous relationships have been set up (e.g. Neely et al. 1974, Kenaga and Goring 1980, Davies and Dobbs 1984). This parameter has been established in decision-making in administrative screening practice as a trigger for requiring higher level testing in the case of new chemicals, or in situations in which data are lacking, for example as a tool for priority setting in the selection of environmentally relevant compounds.

2.3 The Lacking Concept for Assessment - a Gross Deficiency

Whereas increasingly sensitive analytical methods are being developed to detect smaller and even smaller concentrations of xenobiotics in the environment, the overall important issue of how to evaluate existing data is left unattended. The present five OECD Guidelines do not provide any information on how to deal with the generated data. Additionally, depending on the experimental design, for one and the same compound, data ranging over several orders of magnitude have been reported.

Lacking are concepts how to assess the extremely heterogenous pool of data submitted:

* Is a numerical classification of BCFs feasible? Is such an endeavour scientifically justified or solely a practical, i.e. pragmatical necessity for decision making?
* Consequently is bioaccumulation ranking possible?

* How, regarding their ecotoxicological relevance, can data efficiently be integrated into the overall concept of environmental hazard assessment and risk analysis?

Since until now generally accepted threshold values for bioaccumulation do not exist, the evidence of bioconcentration is "merely" regarded as a qualitative reference to risk enhancement. In order to develop criteria for a quantitative assessment, the Federal Environmental Agency has initiated and sponsored a R & D-project, entitled "Entwicklung von Kriterien zur Bewertung der Bioakkumulation für den Vollzug ChemG/PflSchG"[+], comprising a literature study, in which existing assessment approaches are compiled, and the present workshop.

2.4 Challenges of the Workshop

The major pre-requisite to a competent quantitative evaluation of bioaccumulation data of a chemical is the comprehensive knowledge and understanding of the key-factors which might become critical when at a screening level estimating the potential to accumulate, or when actually performing bioaccumulation studies. As quality and appropriateness of submitted data give rise to several problems, mainly regarding their validity, comparability, transferability as well as ecotoxicological relevance, these points therefore represent the major topics of this workshop. Hence, the speakers and the audience are challenged to interact on the following three issue-complexes:

A. Validation

Do the present test guidelines yield reliable results and are they sufficient? How can modelling of bioconcentration be performed? How accurate are log P_{ow} - log BCF - correlations?

B. Extrapolation

Can test results be extrapolated between organism groups? Can laboratory results be extrapolated to environmental conditions and vice-versa? Is a distinction between bioconcentration and biomagnification necessary?

C. Evaluation

Does bioaccumulation reflect toxicity? How can data on bioaccumulation differentially be evaluated? Which current approaches to assess bioaccumulation exist and are they considered adequate to undergo further development?

+ "Development of criteria for the evaluation of bioaccumulation within the execution of ChemG/PflSchG"

As initially pointed out, the overall objective of this workshop is to obtain generalizing statements and to develop a comprehensive evaluation program for bioaccumulation. This should also include a conceptional approach to interspecies as well as lab-to-field extrapolation.

2.5 References

Borgmann, U., Norwood, W.P. and K.M. Ralph, (1990). Chronic Toxicity and Bioaccumulation of 2,5,2',5'- and 3,4,3',4'- Tetrachlorobiphenyl and Aroclor 1242 in the Amphipod Hyalella azteca, *Arch. Environ. Contam. Toxicol.*, 19, 558.

Broekhuizen, S. and E.M. De Ruiter-Dijkman (1988). Otters Lutra lutra with PCBs: Freshwater Seals?, *Lutra*, 31, 68.

Davies, R.P. and A.J. Dobbs, (1984). The Prediction of Bioconcentration in Fish, *Water Res.*, 18, 1253.

Frische, R., Klöpffer, W. and W. Schönborn (1979). Bewertung von organisch-chemischen Stoffen und Produkten in Bezug auf ihr Umweltverhalten - chemische, biologische und wirtschaftliche Aspekte, Umweltbundesamt, Berlin.

Hawker, D.W., (1990). Bioaccumulation of Metallic Substances and Organometallic Compounds, in *Bioaccumulation of Xenobiotic Compounds*, Connell, D.W., Ed., CRC Press, Boca Raton, p. 187.

Hoffman, D.J., Rattner, B.A. and R.J. Hall, (1990). Wildlife Toxicology, *Environ. Sci. Technol.*, 24, 276.

Hose, J.E., Barlow, L.A., Bent, S., Elseewi, A.A., Cliath, M., Resketo, M. and C. Doyle, (1986). Evaluation of Acute Bioassays for Assessing Toxicity of Polychlorinated Biphenyl-contaminated Soils, *Regul. Toxicol. Pharmacol.*, 6, 11.

Kenaga, E.E., (1972). Chlorinated Hydrocarbon Insecticides in the Environment: Factors Related to Bioconcentration of Pesticides, in *Environmental Toxicology of Pesticides*, Matsamura F., Boush, G.M. and T. Misato, Eds., Academic Press, New York, p. 193.

Kenaga, E.E. and Goring, C.A.I., (1980). Relationship between Water Solubility, Soil Sorption, Octanol Water Partitioning, and Concentration of Chemicals in Biota, in *Aquatic Toxicology*, Eaton, J.G., Parrish, P.R. and A.C. Hendricks, Eds., Vol. 707, American Society for Testing and Materials, Philadelphia, p. 78.

Neely, W.B., Branson, D.R. and Blau, G.E., (1974). Partition Coefficient to Measure Bioconcentration Potential of Organic Chemicals in Fish, *Environ. Sci. Technol.*, 8, 1113.

Walker, C.H., (1990). Kinetic Models to Predict Bioaccumulation of Pollutants, *Funct. Ecol.*, 4, 295.

BCF and P: Limitations of the Determination Methods and Interpretation of Data in the Case of Organic Colorants

Peter Moser and Rudolf Anliker

3.1 Abstract

Methods for measurement and estimation of octanol/water partition coefficients are discussed with particular reference to different classes of organic colorants (ionic dyes, disperse dyes, and pigments). Bioaccumulation testing and assessment of BCF results encounter particular problems when the aqueous solubility of the test compound is in the range or lower than the prescribed concentration in the fish test water. Very few accumulating colorants have yet been reported for three reasons: either they are too hydrophilic, their molecular size is too large, or their water solubility is too low. The decision scheme proposed earlier for bioaccumulation screening for organic colorants has been found adequate in all cases investigated so far.

3.2 Introduction

Synthetic organic colorants are not readily biodegradable (Brown and Hamburger 1987) and it is possible that they may enter the aqueous environment, although the amounts involved are relatively small and they are steadily being reduced by improved application technology and effluent treatment. An assessment of their possible environmental impact is nevertheless necessary (Brown 1987) and this involves a consideration of their bioaccumulation tendency.

The partition coefficient n-octanol/water, log P, has become one of the most important parameters in studies on the the fate and impact of organic chemicals in the environment. It has been used to predict sorption and mobility in sediments (Karickhoff 1981, Brown and Flagg 1981), it is one of the main variables in QSAR models for fish toxicity (Könemann

1981, Lipnick et al. 1987, Veith and Broderius 1987; Fiedler et al. 1990) and it has been particularly useful in the prediction of bioaccumulation in fish (for a recent review see Barron 1990). Some of the linear regression equations developed by different laboratories, using different species of fish, different structural classes and different preparations (whole fish vs. body fat) have recently been reviewed by Esser (1986) and by Schüürmann and Klein (1988). In the general bioaccumulation equation

$$\log BCF = a \cdot \log P + b \tag{1}$$

slopes between 0.6 and 1 are generally calculated. For polyaromatic hydrocarbons (PAH) and chlorinated hydrocarbons (CHC), slopes near 0.75 were found in the investigations with the largest number of compounds. Intercepts vary more widely between -1.8 and +0.7, with b = -0.3 as a reasonable average value for PAHs and CHCs. With these coefficients a log P = 3 leads to a predicted log BCF of 1.95. However, Schüürmann and Klein have pointed out that even when validated log P values were used, the log BCF/log P correlations were often not as significant as assumed. Part of this was ascribed to the varying accuracy of the log BCF values.

Environmental regulations requiring ecotoxicological data of newly produced chemicals have been established in most industrialized countries (Caspers and Schüürmann 1991). The Japanese Chemical Substances Control Law (Japan 1973) was the first piece of legislation requiring a bioaccumulation test for the notification of new substances. These tests are mandatory for nonionic substances which (a) are not readily biodegradable and (b) whose n-octanol/water partition coefficient is greater than 1000 (log P > 3). The method for testing the degree of bioaccumulation, expressed as the bioconcentration factor, BCF, was actually the basis for the OECD Test Guideline 305 C (OECD 1981).

This regulation has caused a substantial testing effort for the innovative dye manufacturing industry because the costly flow-through fish tests had to be carried out for a large fraction of the notified dyes. The reasons were, on the one hand, that the coloring agents proved to be not well degradable and thus did not fulfill the first criterion and, on the other hand, a sizable proportion of non-ionic lipophilic disperse, vat and solvent dyes had a log P > 3. The practically insoluble pigments had, almost without exception, values of log P > > 3.

The large group of very hydrophilic cationic and anionic dyes are strongly dissociating and do not fit this scheme. Even though these dyes have very low log P values at ambient pH the Japanese regulations still call for fish accumulation studies as discussed later.

All ionogenic and all strongly hydrophobic dyes, as well as the pigments which were tested according to the Japanese protocol prior to 1981 showed no or only very marginal bioaccumulation (Anliker, Clarke and Moser 1981; ETAD 1990). These results led to the idea that for coloring agents with much more complex chemical structures than ordinary chemicals of ecological concern, log P alone is not sufficient to judge the bioaccumulation tendency. Additional factors like affinity to substrate, molecular size and flexibility, and particularly in the case of organic pigments, the extremely low solubilities in water and body fat must be considered as well. On the basis of a large body of experimental material, we have tried to formulate criteria which consider in a differentiating way the specific properties of the individual classes of coloring agents, and to use these criteria as guidance for deciding on the necessity of performing the bioaccumulation testing.

In a series of papers we have investigated the particular properties of the different classes of coloring agents: At the lower end of the lipophilicity scale we discussed the ionic acids, direct, basic and reactive dyes, most of which have log P values far below zero (Anliker et al. 1981). For this class we have found that it can be confidently predicted that if their log P is $<< 3$ or their water solubility is > 2 g/l then the bioconcentration factor in fish will be < 100 (Anliker et al. 1981).

At the opposite end of the lipophilicity scale, i.e. in the range of extremely high partition coefficients, it was demonstrated (Anliker and Moser 1987) that even if a high log P would predict a strong bioaccumulation tendency, many very lipophilic dyestuffs and pigments do in fact not accumulate in fish under the conditions tested. One explanation for this behaviour was that such compounds, and in particular the pigments, have extremely low water and fat solubilities and thus a very low free concentration in water and a very low fat storage potential. It was calculated that for a typical pigment with a log P = 8.1 and a water solubility in the range of $2\text{-}10 \cdot 10^{-6}$ppm one would predict a lipid concentration of 250 to 1200 ppm. However, the measured lipid (n-octanol) solubility is only 0.46 ppm. Thus, based on a typical lipid content of fish of 5 %, one would in fact expect a pigment content of not more than 0.023 ppm for the whole fish.

For disperse dyes, which form one of the largest single groups of dyes in the textile industry, we have found (Anliker et al. 1988) that even when their log P is > 3 and their solubility is not negligible, they do not bioaccumulate in fish when their molecular size is large presumably because inhibited membrane permeability prevents an effective uptake during the time of exposure (Gobas et al. 1986).

As a result of all of these considerations we have proposed and discussed a decision scheme, shown in Figure 3.1 (Anliker et al. 1988)

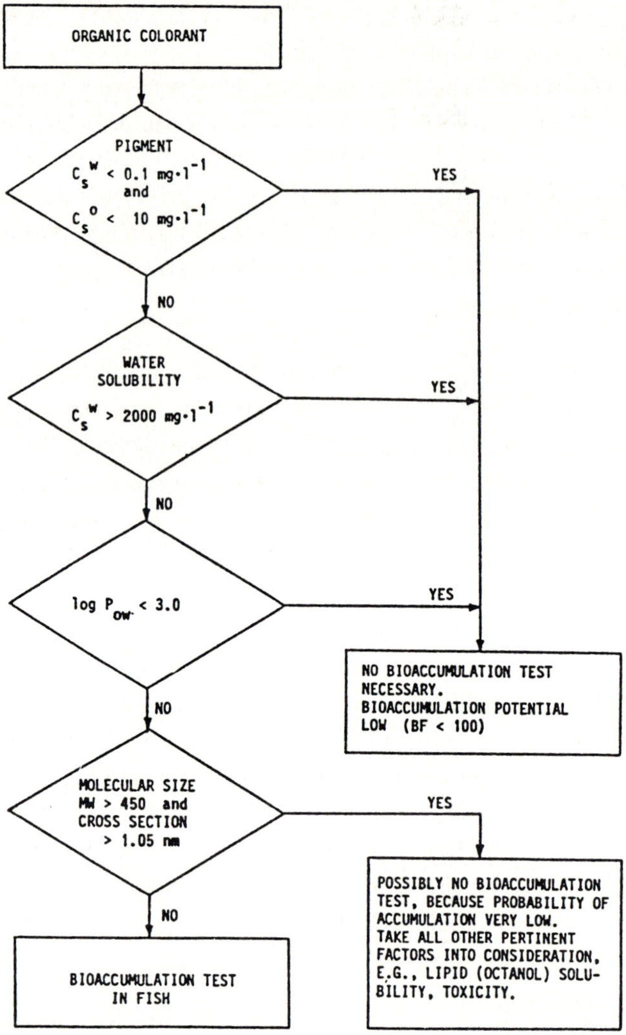

Figure 3.1 Decision Scheme for Bioaccumulation Screening of Organic Colorants (Ionic and Non-ionic Dyestuffs, and Pigments). From Anliker et al. (1988).

In this decision scheme, conservative limits have been chosen to take into account all possible uncertainties. For example, the limiting value of 2000 mg/l for the water solubility, C_S^W, proposed by the OECD Expert Group (1979) was maintained although it was proposed by van Gestel et al. (1985) that a limiting value of 3 mMol/l would be more appropriate. Also, one of the molecular size parameters, 'cross section', (second smallest v. d. Waals diameter) was chosen as > 1.05 nm, which is larger by 0.1 nm than the upper value limiting bioaccumulation suggested by Opperhuizen et al. (1985).

In the course of our work with coloring agents we have encountered several problems and interesting results, some of which we want to discuss in this contribution because we consider them of interest in a wider context.

3.3 Bioaccumulation Testing

A critical appraisal of the different methods for testing for bioconcentration is beyond the scope of this work. They are discussed e.g. in connection with the methods 305 A to E of the OECD Test Guidelines (OECD 1981), and recently by Caspers and Schüürmann (1991). In this section we discuss particularly some aspects of the difficulties encountered when testing dyestuffs and organic pigments, and interpreting results obtained with the Japanese method (OECD method 305 C). Comments on this method, including experience with testing of colorants were given by Hamburger (1980).

In general, the greatest difficulties experienced in testing colorants were problems involving bioavailability (Barron 1990) and analytics. This involved (a) the preparation of stock and test solutions, or rather dispersions, in the case of the scarcely soluble lipophilic dyes and in the case of the even more lipophilic organic pigments, and (b) the reliable determination of the colorants in fish at levels below 1 ppm. In most cases, gas chromatographic methods were not applicable and the less sensitive and less specific spectrophotometric methods had to be used.

With the ionogenic colorants difficulties were only encountered, when, for toxicity reasons, the lower test concentration had to be set at less than 0.1 mg/l. In such exceptional cases an analytical method ten times less sensitive than the lower test concentration was accepted. This meant, however, that only BCF values higher than 10 could be determined. If it was suspected that reactive dyes would chemically react with fish tissue, then the fish material was degraded with proteolytic enzymes in order to isolate the dye bound to protein. As prescribed in the official method the pH of the test solutions were adjusted to 7.0 ± 1.0.

With the sparsely soluble disperse dyes the two test concentrations had, as a rule, to be set at 0.01 and 0.1 mg/l or 0.1 and 1.0 mg/l, resp. Dye concentrations in the fish resulting from lower bath concentrations would have caused an inappropriate analytical effort. For most of the 23 disperse dyes tested (Anliker, Clarke and Moser 1981) the test concentrations were in fact higher than the aqueous solubilities. Yen et al. (1989) have recently been able to confirm this by their measurements of the water solubilities of a number of disperse dyes by the generator column method. The measured water solubilities of Yen et al. (1989) were found lower by factors 2 to 30'000 than the solubilities estimated in our work (Anliker and Moser 1987), which were calculated from log P, melting point and entropy of fusion, according to Yalkowsky and Valvani (1980). The scepticism against such calculated values, which are also used by other authors, seems to be warranted in the case of the complex disperse dyes.

The solubilities of the 10 compounds measured by Yen et al. (1989) were between 0.0001 and 0.276 mg/l. As already remarked, this indicates that most of these dyes were actually tested for BCF above their solubilities. This prompted these authors to assume that, if tested at the concentrations of true solubility, the BCF might be much higher than reported by Anliker et al. (1981). However, it should also be noted that some of the disperse dyes reported (Anliker et al. 1981) had indeed been tested in the range of their water solubilities, e.g. dye No. 2 (C_S^w = 0.0745 mg/l) or dye No. 9 (C_S^w = 0.276 mg/l). The experimental BCF's of both dyes were less than 10 although the calculated log P's were > 3.

Nevertheless, it is possible that not all BCF values of those dyes with C_S^w < 0.01 to 0.1 mg/l are based on truly dissolved organic dye. It is to be expected instead that for some of these dyes the experimental BCF values correspond to the dye concentration measured in fish divided by the total concentration of dye applied to the test system. This total may include truly dissolved dye, aggregated dye, dye adsorbed on organic matter of the test water, dye solubilized by dispersing agents (Jones 1984) and dye contained in the condensed phase. However, the experience gained so far suggests that the lower apparent BCF values caused by these effects may be partially compensated or even surpassed by dye adsorbed to the exterior of the fish or to particulate dye material adhering to organs like the gills or the stomach. In addition it is known that disperse dyes have a tendency to form supersaturated solutions in the concentration range of testing; thus concentrations in excess of C_S^w are expected in the test water.

A very critical problem is the preparation of stable stock solutions of dyes and pigments with very low solubility. The OECD method 305 C proposes the use of several solvents and surfactants for solubilizing substances with such characteristics. The preferred surfactants in the tests reported by Anliker at al. (1981) were Tween 80 and Hardened Castor Oil 40. The Japanese experts are of the opinion that the use of these proposed surfactants (OECD method 305 C, 1981) at a ratio not exceeding 1:10 (test compound: surfactant) should not have a noticeable influence on the BCF. However, since systematic investigations are not available, there is still no consensus as to the absence of any influence on the BCF. Even if the discussed limitations of the test method are taken into account, the overall BCF values of more than 80 disperse dyes indicate that in general this class shows a lower tendency to bioaccumulate than would be expected based on the log P alone. It would certainly be desirable to test those disperse dyes with very low solubilities at test concentrations below or at saturation, i.e. in the range of 0.0001 to 0.01 mg/l. But this will require the use of radiolabelled substances or application of new, more sensitive analytical procedures for the determination of the dyes in fish.

Testing of the organic pigments which not only have an extremely low solubility in water but also in fat (octanol) becomes even more problematic. These solubility characteristics make a sensible experimental determination of the BCF practically impossible. Therefore, one has to rely on other parameters for a realistic estimate of the bioaccumulation potential of pigments as outlined in the decision scheme shown in Figure 3.1.

3.4 Partition Coefficients

Schüürmann and Klein (1988) have pointed out the great importance of using partition coefficients of high quality for QSARs predicting bioaccumulation. It is obvious from equation (1) that an error of one power of ten in log P results in an error of almost one power of ten in the predicted BCF.

For the dyestuffs the situation is maybe even more difficult than for the 'ordinary' environmental chemicals (pesticides, halogenated hydrocarbons) because they span an even wider lipophilicity range - from the very hydrophilic ionic dyes to the superlipophilic, practically insoluble, pigments - and they are often not available for measurement in purified form. In each class of coloring agents we have found some particular problems:

The range $-2.5 < \log P < 7$

The shake flask method which is included in the original OECD Test Guidelines (1981) is the classical procedure for the determination of P. It is most suitable in the range $-2.5 < \log P < 4$ but can, by using very small volume ratios of n-octanol to water, be applied up to $\log P = 6$ (USEPA 1988). It has been used whenever possible in our own work (Anliker et al. 1981, 1987, 1988). One of its main limitations, as discussed by Esser and Moser (1982), is its sensitivity to impurities, particularly at the extremes of the log P range. This problem can, however, be partly overcome by substance specific concentration determination, e.g. by HPLC, GC or MS.

Last year, the much awaited OECD Test Guideline 117 on HPLC has finally appeared (OECD 1989), which is based on work initiated by ECETOC (Eadsforth and Moser 1983) and which was ring tested and brought to the current standard under the guidance of the German

Environmental Protection Agency (UBA) (for recent reviews on the RP-HPLC method, see Unger et al. 1986; Kaliszan 1990). Although the HPLC method is in principle applicable to hydrophobic disperse dyes, reliable log P values are difficult to obtain due to the lack of validated reference compounds (see Test Guideline 117) resembling ionic dyes or dyes in the highly lipophilic range. No literature citations have been found, though, for the use of HPLC for predicting bioaccumulation of dyes by the regression method.

According to our experience, calculation of log P by the fragment procedure of Hansch and Leo, using the program CLOGP (Pomona 1986), yields reliable values when applied properly, i.e. when all possible effects, which are either not or incorrectly parametrized in the computer program are considered. Mainly, effects like intramolecular hydrogen bonds, odd fragments or partial dissociation must be corrected manually.

Baughman and Perenich (1988) and Yen et al. (1989) reported consistently lower experimental log P values for 10 disperse dyes than the ones calculated from their own experimental water solubilities using the scheme of Yalkowsky and Valvani (1980). The substituent approach CLOGP yields even lower log P's. Part of these discrepancies were ascribed to an underestimation of solvent interaction, aggregation of the disperse dyes and/or experimental errors. The authors judge their experimental water solubilities and log P's derived thereof as the most trustworthy and consider measured or calculated log P's rather as lower limits for the estimation of bioaccumulation. From their data on water solubility these authors predict log BCF's between 3 and 5 for their 10 dyes. High bioaccumulation is, however, in contrast with the actual experience (Anliker et al. 1988; ETAD 1990). Figure 3.2 shows that from 65 lipophilic disperse dyes tested (ETAD 1990), bioaccumulation with BCF > 100 was only observed with 2 dyes. Molecular sizes of those dyes are small (MW < 450), i.e. they are in the range where uptake through biological membranes seems to be easier. This had already been correctly considered in our decision scheme, Figure 3.1. All other 63 disperse dyes, even those with MW < 450, showed BCF < 100.

The range of superlipophilic compounds

A number of papers have recently appeared describing methods for reliably measuring partition coefficients in the range P > 105, using either the generator column (Wasik et al. 1983; Doucette and Andren 1988, Shiu et al. 1988) or the slow stirring technique (de Bruijn et al. 1989, Brooke et al. 1990). With PAH's and CHC's it was possible to attain log P's of more than 8 (decachlorobiphenyl, log P = 8.27, de Bruijn et al.) and to settle the dispute on the log P of the important pesticide p,p-DDT: the value of log P = 6.91, measured by de Bruijn et al., corresponds exactly to the value calculated by CLOGP (Anliker and Moser 1987).

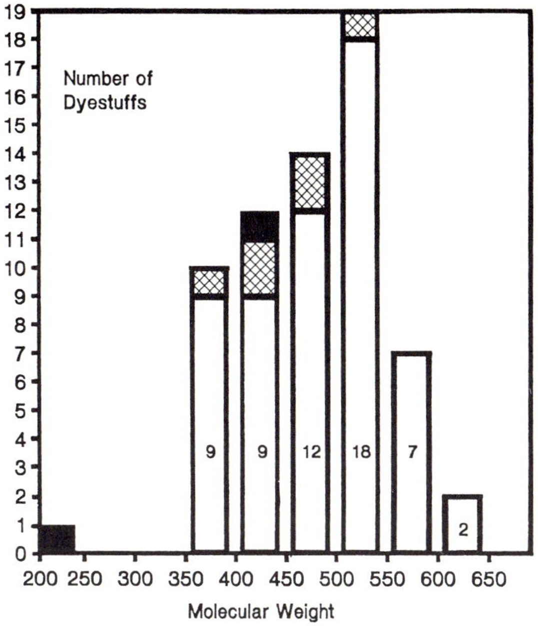

Figure 3.2 Relationship Between BCF and Molecular Weight for 65 New Disperse Dyestuffs Registered in Japan.
Bar diagram showing the number of dyes with log BCF in the ranges
filled: 2.0 s log BCF < 3.4
hatched: 1.5 s log BCF < 2.0
blank: log BCF < 1.5
Published with kind permission of Japanese ETAD member companies.

The superlipophilic compounds of interest to the dye manufacturers are the pigments. However, while even the extremely lipophilic dibenzo-p-dioxins have water solubilities in the range of 0.1 to 100 μg/l (Shiu et al. 1988), most of the organic pigments are soluble in water to less than 10^{-3} μg/l (Anliker and Moser 1987). In addition they are also almost insoluble in organic solvents (C_S in n-octanol is in the mg/l range) making it virtually impossible to determine meaningful partition coefficients experimentally.

The ionized compounds

Most ionic dyes are rendered hydrophilic by incorporating one or several $SO3^-$- or quaternary ammonium groups. Each such group contributes about -5 log units to hydrophilicity of the total molecule. Thus, such ionic dyes usually range from rather hydrophilic to extremely hydrophilic with (calculated) log P values down to unrealistically low -10 to -13 (Anliker et al. 1981).

The sulfonic acid group is strongly acidic, with a $pK_a < 1$. In an earlier paper (Esser and Moser 1982) it was shown, that the log P of a representative model compound, naphthalene sulfonic acid, remains practically constant over a wide pH-region. In Fig. 3.3a we have extended the measured pH range down to 0, we have measured at two buffer strengths, and we have added the calculated value for the undissociated molecule. One can recognize a log P/pH profile which is typical for all acids (Moser et al. 1990): a plateau region below the pK_a, representing the unionized free acid (log P_u), followed by a sigmoidal transition into an other plateau region representing the log P of the fully ionized compound. In the region of partial dissociation (between pH -1 and pH +2 in this case), P relates to P_u by equation (2) which describes the situation (Moser et al. 1990) that only the fraction of unionized species partitions between the aqueous and lipid phase.

$$\log P = \log P_u - \log(1+10^{pH-pK}) \qquad (2)$$

The inflection point suggests a pK_a of -0.67 for the 1-naphthalene sulfonic acid. In the presence of counterions, either in the form of buffer ions (Figure 3.3a) or as added neutral salts (Figure 3.3b), ion pair partitioning is observed. At physiological sodium chloride concentration (0.9 g/100 ml) this effect may increase log P of an ionized dye molecule by 0.5 to 1 orders of magnitude.

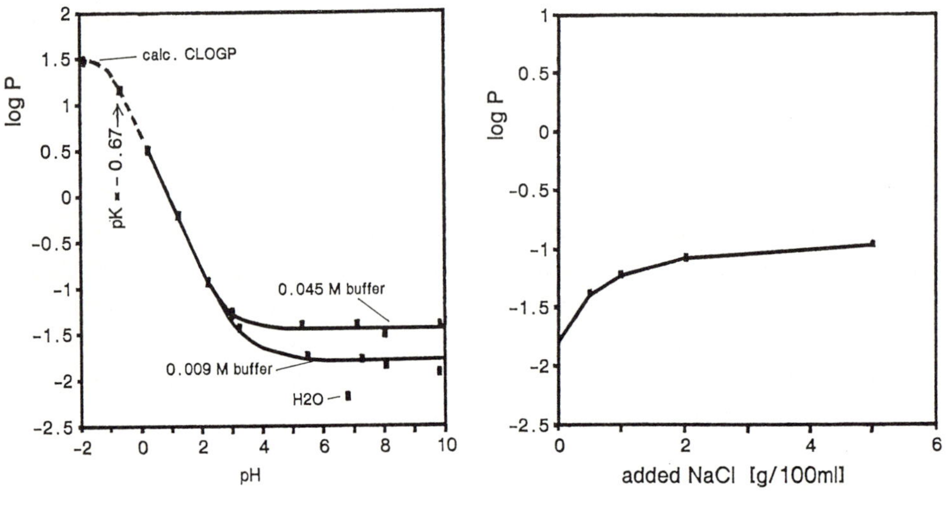

Figure 3.3 Log P (Partition Coefficient n-Octanol/Water) of 1-Naphthalene Sulfonic Acid.
a) pH-dependence, measured at two different buffer concentrations and in dest. water. The point at pH -2 is the calculated CLOGP value, the inflection point corresponds to a pK_a of -0.67.
b) effect of added neutral salt (in 0.009 M phosphate buffer, pH 7.4).

In a recent paper (Wellenreuther 1989) attention was drawn to the fact that the Japanese Ministry of International Trade and Industry (MITI) has introduced an additional new prerequisite for performing the bioaccumulation test OECD 305 C, concerning ionizable compounds. According to the Chemical Substances Control Law of Japan, a bioaccumulation test in fish can be substituted by the determination of the partition coefficient n-octanol/water as long as the three conditions are satisfied:

1. the substance is not insoluble in water
2. the substance should not show dissociation or association in water
3. the substance should not be surface active.

According to MITI a given substance is said to dissociate in water when the following conditions are satisfied:

for acids: $pKa - pH < 1.7$
for bases: $pH - pKa < 1.7$

where pH is the pH of the aqueous phase for the determination of log P. Applied to our model dye, 1-naphthalene sulfonic acid, this requirement would mean that it would be considered dissociated in the whole pH range above $(-0.67-1.7)=-2.37$. Thus, a bioaccumulation test would have to be performed, even if the log P is more than 4 powers of ten lower than log P = 3 over the whole accessible and sensible pH range!

As an alternative, we propose to measure log P at 3 different pH values spanning the pH-range possibly occurring in surface waters, even under adverse conditions: pH 5, pH 7 and pH 9. These three pH values prescribed in the EPA testing guideline on partition coefficients (USEPA, 1988) would appear adequate for this purpose. If log P is < 3 at all three pH values, then a bioaccumulation test is not necessary. Our proposal is based on the well known fact (Scherrer and Howard 1977; Martin 1978) that in QSAR's involving transport of dissociating substances through or into biological phases (membranes, fat), the log P of the partly or fully dissociated compound (also called distribution coefficient, D) is the governing factor, and not the partition coefficient of the undissociated compound, log P_u.

3.5 Conclusions

The n-octanol/water partition coefficient, P, is accepted to be a good predictor of the tendency of different chemicals to bioaccumulate in fish and linear correlations between log BCF and log P exist. These allow the estimation of BCF values from a simple determination of P. As most correlation equations are based on data of a restricted range of chemicals, mostly relatively simple halogenated aliphatic and aromatic substances, it could have been expected, that with more complex compounds having strongly different physicochemical features other factors would have to be taken into account.

In connection with the performance of a considerable number of fish bioaccumulation tests with colorants it was recognized that the simple linear QSAR model indeed has its limitations in this heterogeneous class of chemicals which includes a wide variety of complex structures. The available data discussed in this contribution indicate that the numerical value of P only allows in the case of ionized dyes to arrive at a sensible decision whether a fish accumulation test should be performed. With the lipophilic disperse and vat dyes and the superlipophilic pigments consideration of further parameters such as water and lipid solubility, molecular size or molecular weight provide an adequate basis for a reliable decision making using the proposed scheme shown in Figure 3.1.

Its use helps to prevent unnecessary fish bioaccumulation tests which are in the case of the highly lipophilic, scarcely soluble disperse dyes linked with considerable experimental difficulties, and in the case of the practically insoluble pigments utterly senseless. As the ionic dyestuffs represent about 45 to 50 % of the total market, their inclusion in the decision scheme is of upmost importance. New regulations by the Japanese authorities which require accumulation tests of all dissociating chemicals, irrespective of their effective log P at the pH of natural waters must be carefully discussed.

3.6 References

Anliker, R., Clarke, E.A. and P. Moser, (1981). Use of the Partition Coefficient as an Indicator of Bioaccumulation Tendency of Dyestuffs in Fish, *Chemosphere*, 10, 263-274.

Anliker, R. and P. Moser, (1987). The Limits of Bioaccumulation of Organic Pigments in Fish: Their Relation to the Partition Coefficient and the Solubility in Water and Octanol. *Ecotoxicol. Environ. Saf.*, 13, 43-52.

Anliker, R., Moser, P. and D. Poppinger, (1988). Bioaccumulation of Dyestuffs and Organic Pigments in Fish. Relationships to Hydrophobicity and Steric Factors. *Chemosphere*, 17, 1631-1644.

Barron, M.G. (1990). Bioconcentration, *Environ. Sci. Technol.*, 24, 1612-1618.

Baughman, G.L. and T.A. Perenich (1988). Fate of Dyes in Aquatic Systems: I. Solubility and Partitioning of some Hydrophobic Dyes and Related Compounds. *Environ. Toxicol. Chem.*, 7, 183-199.

Brooke, D., Nielsen, I., de Brujin, J. and J. Hermens (1990). An Interlaboratory Evaluation of the Stir-Flask Method for the Determination of Octanol-Water Partition Coefficients (Log P_{OW}), *Chemosphere*, 21, 119-133.

Brown, D.S. and E.W. Flagg, (1981). Empirical Prediction of Organic Pollutant Sorption in Natural Sediments, *J. Environ. Qual.*, 10, 382-386.

Brown, D. and B. Hamburger, (1987). The degradation of dyestuffs: Part III. Investigation of their ultimate degradability, *Chemosphere*, 16, 1539-1553.

Brown, D. (1987). Effects of Colorants in the Aquatic Environment, *Ecotoxicol. Environ. Saf.*, 13, 139-147.

Caspers, N. and G. Schüürmann (1991). Bioconcentration of Xenobiotics from the Chemical Industry's Point of View, *These Proceedings*.

de Bruijn, J., Busser, F., Seinen, W., and J. Hermens, (1989). Determination of Octanol/Water Partition Coefficients for Hydrophobic Organic Chemicals with the "Slow-Stirring Method", *Environ. Toxicol. Chem.*, 8, 499-512.

Doucette, W.J. and A.W. Andren, (1988). Estimation of Octanol/Water Partition Coefficients: Evaluation of Six Methods for Highly Hydrophobic Aromatic Hydrocarbons, *Chemosphere*, 17, 345-359.

Eadsforth, C.V., and P. Moser (1983). Assessment of Reverse-Phase Chromatographic Methods for Determining Partition Coefficients. *Chemosphere*, 12, 1459-1475.

Esser, H.O. and P. Moser (1982). An Appraisal of Problems Related to the Measurement and Evaluation of Bioaccumulation, *Ecotoxicol. Environ. Safety*, 6, 131-148.

Esser, H.O. (1986). A Review of the Correlation Between Physicochemical Properties and Bioaccumulation, *Pestic. Sci.*, 17, 265-276.

ETAD (1990). Communication to ETAD by Japanese ETAD member companies on results of bioaccumulation tests on 65 disperse dyes performed according to the Japanese standard test method (OECD method 305 C)

Fiedler, H., Hutzinger, O. and J.P. Giesy (1990). Utility of the QSAR Modeling System for Predicting the Toxicity of Substances on the European Inventory of Existing Commercial Chemicals, *Toxicol. Environ. Chem.*, 28, 167-188.

Gobas, F.A.P.C., Opperhuizen, A. and O. Hutzinger (1986). Bioconcentration of Hydrophobic Chemicals in Fish: Relationship with Membrane Permeation, *Environ. Toxicol. Chem.*, 5, 637-646.

Hamburger, B. (1980). German Experience with the Japanese Fish Accumulation Test, *Ecotox. Environ. Safety*, 4, 17-20.

Japan (1973). Japanese Chemical Substances Control Law (No. 117). Order of the Japanese Prime Minister, the Minister of Health and Welfare, and the Minister of International Trade and Industry, No. 1, Promulgated July 13, 1974.

Jones, F. (1984). The Solubility and Solubilisation of Disperse Dyes, *J. Soc. Dyers & Colourists*, 100, 66-72.

Kaliszan, R. (1990). High Performance Liquid Chromatographic Methods and Procedures of Hydrophbocoty Determination, *Quant. Struct.-Act. Relat.*, 9, 83-87.

Karickhoff, S.W. (1981). Semi-empirical Estimation of Sorption of Hydrophobic Pollutants on Natural Sediments and Soil, *Chemosphere*, 10, 833-846.

Könemann H. (1981). Quantitative Structure-Activity Relationships in Fish Toxicity. Part 1: Relationship for 50 Industrial Pollutants, *Toxicology*, 19, 209-221.

Lipnick, R.L., Watson, K.R. and A.K. Strausz (1987). A QSAR Study of the Acute Toxicity of Some Industrial Organic Chemicals to Goldfish. Narcosis, Electrophile and Proelectrophile Mechanisms, *Xenobiotica*, 17, 1011-1025.

Martin, Y.C. (1978). Quantitative Drug Design. Marcel Dekker Inc., New York & Basel.

Moser, P., Sallmann, A. and I. Wiesenberg (1990). Synthesis and Quantitative Structure-Activity Relationships of Diclofenac Analogues, *J. Med. Chem.*, 33, 2358-2368.

OECD Expert Group on Degradation and Accumulation (1979). Final Report,, Vol.I, Parts 1+2, Umweltbundesamt, FRG; Government of Japan. Berlin and Tokyo, December.

OECD (1981). OECD Guidelines for Testing of Chemicals. 1. Physical-Chemical Properties / 3. Degradation and Accumulation. Bureau of Information, Paris.

OECD (1989). OECD Guidelines for Testing of Chemicals. 1. Physical-Chemical Properties. No. 117. Bureau of Information, Paris.

Opperhuizen, A., v.d.Velde, E.W., Gobas, F.A.P.C. and J.D.M. v.d.Steen, (1985). Relationship between Bioconcentration in Fish and Steric Factors of Hydrophobic Chemicals, *Chemosphere*, 14, 1871-1896.

Pomona College Medicinal Chemistry Project, Claremont, Calif. 91711, Program CLOGP 3.42, (1986).

Scherrer, R.A. and S.M. Howard (1977). Use of Distribution Coefficients in Quantitative Structure-Activity Relationships, *J. Med. Chem.*, 20, 53-58.

Schüürmann G. and W. Klein (1988). Advances in Bioconcentration Prediction, *Chemosphere*, 17, 1551-1574.

Shiu W.Y., Doucette, W., Gobas, F.A.P.C., Andren, A. and D. Mackay (1988). Physical Chemical Properties of Chlorinated Dibenzo-p-dioxins. *Environ. Sci. Technol.*, 22, 651-658.

Unger, S.H., Cheung, P.S., Chiang, G.H. and J.R. Cook (1986). In *Partition Coefficient. Determination and Estimation*, Dunn, W.J., Block J.H., and R.S. Pearlman, Eds., Pergamon Press, New York, pp. 69-81.

USEPA (1988). U.S. Environmental Protection Agency. Chemical Fate Testing Guidelines. 796.1550 Partition Coefficient (n-Octanol/water). 5-13-88.

van Gestel, C.A.M., Otermann, K. and J.H. Canton (1985). Relation between Water Solubility, Octanol/Water Partition Coefficients, and Bioconcentration of Organic Chemicals in Fish: A Review, *Regulat. Toxicol. Pharmacol.*, 5, 422-431.

Veith, G.D. and S.J. Broderius (1987). Structure-Toxicity Relationships for Industrial Chemicals Causing Type (II) Narcosis Syndrome, In *QSAR in Environmental Toxicology - II.* Kaiser, K.L.E., Ed., D. Reidel Publ. Company, pp. 385-391.

Wasik, S.P., Miller, M.M., Tewari, M.M., May, W.E., Sonnefeld, W.J., de Voe, H. and W.H. Zoller (1983). Determination of the Vapor Pressure, Aqueous Solubility, and Octanol/Water Partition Coefficient of Hydrophobic Substances by Coupled Column/Liquid Chromatographic Methods. *Residue Rev.*, 85, 29-42

Wellenreuther, G. (1989). Neue Entwicklungen bei den japanischen Chemikaliengesetzen, *Melliand Textilber.*, 71, 466-470.

Yalkowsky, S.H. and C.S. Valvani (1980). Solubility and Partitioning: I. Solubility of Nonelectrolytes in Water, *J. Pharm. Sci.*, 69, 912-922.

Yen, C.-P.C., Perenich, T.A., and G.L. Baughman (1989). Fate of Dyes in Aquatic Systems II, Solubility and Octanol/Water Partition Coefficients of Disperse Dyes, *Environ. Toxicol. Chem.*, 8, 981-986.

Mathematical Description of Uptake, Accumulation and Elimination of Xenobiotics in a Fish/Water System

Werner Butte

4.1 Calculation of Bioconcentration Factors

In general, there are two different methods to evaluate bioconcentration factors (BCF): one is to calculate it from the concentration of a chemical in fish (c_f) divided by the concentration in water (c_w) under steady state conditions, that means, for an equilibrium partition of the substance between fish and water. The other method is the calculation of BCF's from kinetic data. A summary of the evaluation of BCF values is given in Table 4.1, regarding the test guidelines of the OECD (1981).

Table 4.1 Evaluation of Bioconcentration Factors (BCF) in a Fish/Water System

Fish Test (OECD 305)	experimental expenditure	calculation of BCF	information obtained
Static (D)	small	c_f/c_w	BCF (BCF_{8d})
Sequential Static (A)	small-medium	c_f/c_w	BCF $= f(c_w)\ k_2$
Semi Static (B)	small-medium	c_f/c_w	BCF_{28d}
Flow-Trough (C) - Plateau -	medium	c_f/c_w	BCF (BCF_{56d})
Flow-Through (E) - Kinetic -	high	k_1/k_2	BCF_∞, k_1, k_2

4.2 Models to Describe the Uptake and Clearance of Chemicals by Fish

4.2.1 The Two-Compartment Model

Assuming a two-compartment model (see Figure 4.1) with first order kinetics for the distribution of chemicals between fish and water (Blau et al. 1975), the bioaccumulation factor (BCF) may be calculated, according to test schedule OECD No. 305 E (OECD 1981), as

$$BCF = \lim_{t \to \infty} (c_f / c_w) = k_1 / k_2 \qquad (1)$$

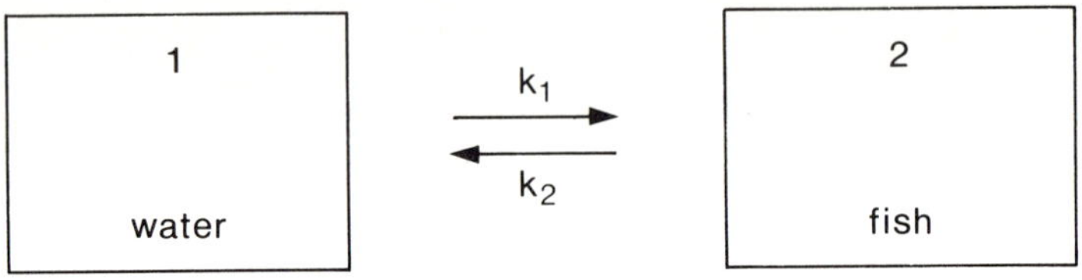

Figure 4.1 Two-Compartment Model for the Distribution of a Chemical between Water and Fish

The differential Equations to describe the uptake and clearance of a chemical in fish using this two-compartment model are:

for the accumulation:

	in water	in fish	
with:	$c_w = c_1 = \text{const.}$	$c_f = c_2$	(2)
	$\dfrac{dc_w}{dt} = 0$	$\dfrac{dc_f}{dt} = k_1 \cdot c_1 - k_2 \cdot c_2$	(3)

and for the clearance:

	in water	in fish	
with:	$c_w = c_1 = 0$	$c_f = c_2$	(4)
	$\dfrac{dc_w}{dt} = 0$	$\dfrac{dc_f}{dt} = - k_2 \cdot c_2$	(5)

4.2.2 The Three-Compartment Model

For the three-compartment model, the fish consists of two separate compartments (see Figure 4.2).

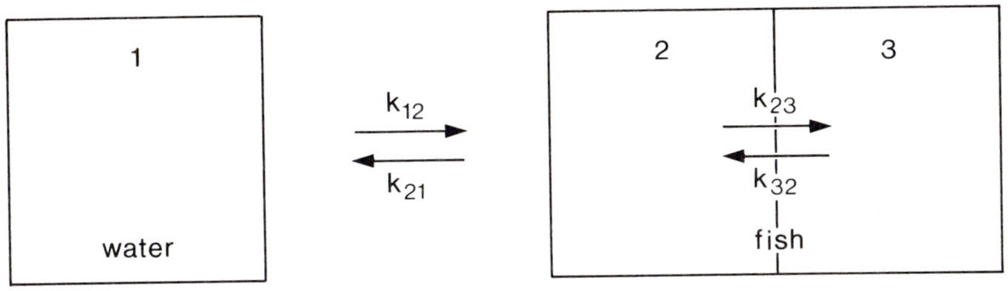

Figure 4.2 Three-Compartment Model for the Distribution of a Chemical between Water and Fish

With first order kinetics for the distribution of a chemical between water and the two fish compartments, the differential Equations to describe uptake and clearance are:

for the accumulation:

<u>in water</u> <u>in fish</u>

with: $c_w = c_1 = \text{const.}$ $$c_f = \frac{m_2 + m_3}{f_w} \qquad (6)$$

$$\frac{dc_w}{dt} = 0 \qquad \frac{dc_2}{dt} = k_{12} \cdot c_1 - k_{21} \cdot c_2 + k_{32} \cdot c_3 \qquad (7)$$

$$\frac{dc_3}{dt} = k_{23} \cdot c_2 - k_{32} \cdot c_3 \qquad (8)$$

for the clearance: <u>in water</u> <u>in fish</u>

with: $c_w = c_1 = 0$ $$c_f = \frac{m_2 + m_3}{f_w} \qquad (9)$$

$$\frac{dc_w}{dt} = 0 \qquad \frac{dc_2}{dt} = - k_{21} \cdot c_2 + k_{32} \cdot c_3 \qquad (10)$$

$$\frac{dc_3}{dt} = k_{23} \cdot c_2 - k_{32} \cdot c_3 \qquad (11)$$

4.2.3 Solutions of the Differential Equations

The Two-Compartment Model

As already reported by Blau et al. (1975) the solutions for the differential equations to describe the variation of the chemical in fish with time under uptake and clearance (depuration) conditions are:

$$c_f = \frac{k_1}{k_2} \cdot c_w \cdot (1 - e^{-k_2 \cdot t}) \tag{12}$$

$$c_f = \frac{k_1}{k_2} \cdot c_w \cdot [e^{-k_2 \cdot (t-t^*)} - e^{-k_2 \cdot t}] \tag{13}$$

The BCF is then:

$$BCF = \frac{k_1}{k_2} \tag{14}$$

The Three-Compartment Model

According to Könemann and van Leeuwen (1980) the accumulation of a chemical for the accumulation phase can be described as:

$$c_f = A \cdot (1 - e^{-\alpha \cdot t}) + B \cdot (1 - e^{-\beta \cdot t}) \tag{15}$$

and for the clearance phase, if the elimination starts under steady state conditions and with a new time scale, beginning with the clearance phase, i.e. te = 0:

$$c_f = A \cdot e^{-\alpha \cdot te} + B \cdot e^{-\beta \cdot te} \tag{16}$$

Without reaching the steady state conditions during the accumulation phase, thus not reaching a steady state and not knowing the concentration of a chemical in fish under equilibrium for the fish/water system - applying for example test No. OECD 305 E - the clearance may be described as:

$$c_f = A \cdot [e^{-\alpha \cdot (t-t^*)} - e^{-\alpha \cdot t}] + B \cdot [e^{-\beta \cdot (t-t^*)} - e^{-\beta \cdot t}] \tag{17}$$

with t* as the time, when the clearance phase starts.

After fitting the data points (t_i, c_{fi}) using non linear regression, thus estimating A, B, α, ß the BCF may be calculated as:

$$BCF = \frac{A + B}{c_w} \tag{18}$$

which means: (A + B) = $c_{f,\infty}$ for t->∞; but it should be noted that A + B don't have a real meaning, thus (A + B) = $c_{f,\infty}$ holds only for t->∞ !

After non-linear regression analysis (see Chapter 4.3.3), thus after obtaining A, B, α, and ß, the kinetic parameters may be calculated according to Könemann and van Leeuwen (1980) as:

$$k_{12} = \frac{\alpha \cdot A + \beta \cdot B}{c_w} \tag{19}$$

$$k_{21} = \alpha + \beta - \frac{\alpha \cdot \beta}{k_{12}} \cdot \frac{A + B}{c_w} = \alpha + \beta - \frac{\alpha \cdot \beta \cdot BCF}{k_{12}} \tag{20}$$

$$k_{32} = \frac{\alpha \cdot \beta}{k_{21}} \tag{21}$$

$$k_{23} = \alpha + \beta - k_{21} - k_{32} \tag{22}$$

4.3 Mathematical Modelling

4.3.1 Graphical Methods

With the two-compartment model the concentration of a compound in fish for the clearance phase may be described as:

$$c_f = C \cdot e^{-k_2 \cdot t} \tag{23}$$

Logarithmic transforming results in:

$$\ln c_f = \ln C - k_2 \cdot t \tag{24}$$

Thus from two measurements of the concentration of a compound in fish under clearance conditions one may calculate k_2 (see Figure 4.3):

$$\ln c_{f1} = \ln C - k_2 \cdot t_1 \tag{25a}$$

$$\ln c_{f2} = \ln C - k_2 \cdot t_2 \tag{25b}$$

Equation (25a) and (25b) give:

$$k_2 = \frac{\ln c_{f1} - \ln c_{f2}}{t_2 - t_1} \tag{26}$$

If the clearance starts under steady state conditions the constant C is identical to:

$$C = \frac{k_1}{k_2} \cdot c_W \tag{27}$$

As k_1/k_2 = BCF [see Equation (14)], C divided by the concentration of the compound in water (C_W), results in the BCF. As k_2 and the BCF are known now, k_1 may be calculated from Equation (14). All this may be done using a graph-paper (semi-log scale) and a pocket calculator.

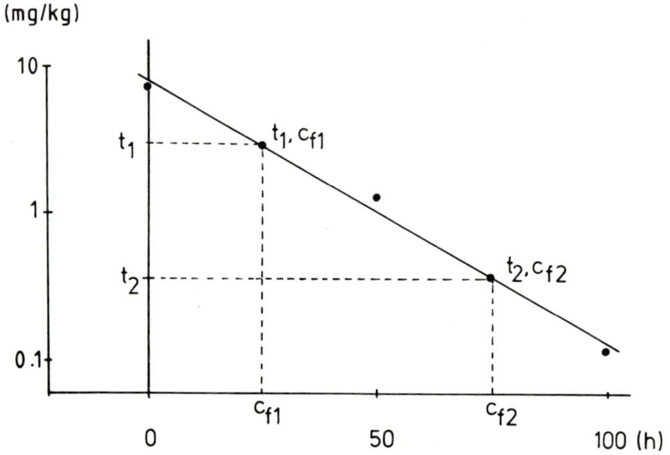

Figure 4.3 Clearance of Pentachlorophenol from Fish

If the clearance phase does not start under steady state conditions, e.g. if OECD test No. 305 E has been used, one has to calculate k_1 according to (OECD, 1981):

$$k_1 = \frac{c_f \cdot k_2}{c_w \cdot (1 - e^{-k_2 \cdot t})} \tag{28}$$

taking a concentration (c_f) and a time (t) from the midpoint of the uptake phase. For the three-compartment model graphical methods may also be applied, but it is more advisable to use linear regression methods as described in the next chapter.

4.3.2 Linear Regression (Least Squares)

The principles of linear regression analysis are described elsewhere (see for example: Montgomery and Peck 1982). If one wants to apply linear regression analysis to estimate k_2 from the clearance phase (two-compartment model) one might use Equation (24), which, for several data points, leads to:

$$\ln c_{fi} = \ln C - k_2 \cdot t_i \tag{29}$$

Least squares estimators are thus obtained for k_2 and lnC. The BCF and k_1 are obtained as described in Chapter 4.3.1. If the three-compartment model seems to be a better model to describe the data of the clearance phase a technique called "feathering" may be used (Knorre 1981).

The variation of the concentration of a chemical in fish during the clearance phase is:

$$c_{fi} = A \cdot e^{-\alpha \cdot t_i} + B \cdot e^{-\beta \cdot t_i} \tag{16}$$

this may be transformed to (mathematically incorrect !):

$$\ln c_{fi} = \underbrace{\ln A - \alpha \cdot t_i}_{Y_{II}} + \underbrace{\ln B - \beta \cdot t_i}_{Y_I} \tag{30}$$

The variation of the concentration of a chemical in fish may thus be regarded to be dependent from two linear equations: Y_I and Y_{II}. Taking the last values of the curve (in a logarithmic scale), they are described by:

$$Y_I = \ln B - \beta \cdot t \tag{31}$$

B and ß are obtained from linear regression analysis. Now

$$Y_{II} = A - \alpha \cdot t_i = c_{fi} - (B - ß \cdot t_i) \quad (32)$$

is calculated. Linear regression leads to A and α (see Figure 4.4).

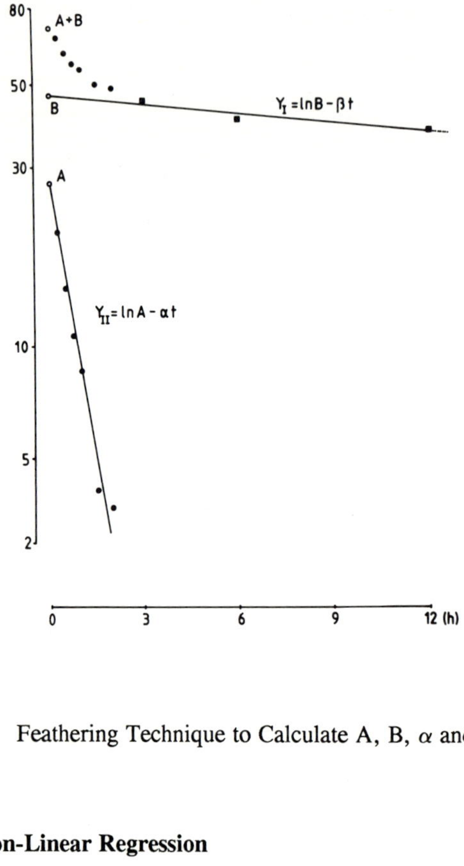

Figure 4.4 Feathering Technique to Calculate A, B, α and ß.

4.3.3 Non-Linear Regression

In Chapter 4.3.2 the application of linear models to obtain the rate constants and the BCF from a bioaccumulation experiment was described, but one always had to divide the experimental data into a accumulation and a clearance phase, never using all data of an experiment simultaneously.

This drawback can be overcome by non-linear regression analysis. A further advantage is, that it is not necessary to obtain a steady state for the accumulation experiment, which is especially useful if highly lipophilic compounds are under evaluation.

Non-linear regression to estimate k_1, k_2 and the BCF for the two-compartment model has been described (Butte and Blum 1984). The principles of the estimation of four parameters by non-linear regres analysis (Gauss-Newton iteration) have been published by Krüger-Thiemer (1964). This approach might be used to estimate A, B, α and ß of Equation (15) and (17).

Taking only the clearance phase of the three-compartment model the concentration of a chemical in fish is (see Equation 16):

$$C_{fi} = A \cdot e^{-\alpha \cdot t_i} + B \cdot e^{-\beta \cdot t_i} = f(A,B,\alpha,\beta) \qquad (16)$$

In non-linear regression, start values for A, B, α and ß are estimated (for example from feathering), i.e. A^o, B^o, α^o, and β^o, and corrections for the start values (ΔA, ΔB, $\Delta\alpha$, $\Delta\beta$) are defined as:

$$A = A^o + \Delta A \qquad (33a)$$
$$B = B^o + \Delta B \qquad (33b)$$
$$\alpha = \alpha^o + \Delta\alpha \qquad (33c)$$
$$\beta = \beta^o + \Delta\beta \qquad (33d)$$

The non-linear function f (A,B,α,ß) may thus be transformed to a linear function g(ΔA, ΔB, $\Delta\alpha$, $\Delta\beta$) by a Taylor series, cutting after the linear term:

$f(A,B,\alpha,\beta) =$

$$f(A^o, B^o, \alpha^o, \beta^o) + \frac{\delta f}{\delta A} \cdot \Delta A + \frac{\delta f}{\delta B} \cdot \Delta B + \frac{\delta f}{\delta \alpha} \cdot \Delta\alpha + \frac{\delta f}{\delta \beta} \cdot \Delta\beta \qquad (34)$$

Assuming, that the experimental concentrations c_{f1}, c_{f2} ... C_{fn} (n = number of measured concentrations) deviate from the theoretical values $f_1(A,B,\alpha,\beta)$, $f_2(A,B,\alpha,\beta)$ .... $f_n(A,B,\alpha,\beta)$ by r_1, r_2 .... r_n the following equations are defined:

$$A_i^o = c_{fi} - f(A,B,\alpha,\beta) \qquad (35a)$$
$$A_i^1 = \delta f/\delta A \qquad (35b)$$
$$A_i^2 = \delta f/\delta B \qquad (35c)$$
$$A_i^3 = \delta f/\delta \alpha \qquad (35d)$$
$$A_i^4 = \delta f/\delta \beta \qquad (35e)$$

and the following error equations are obtained:

$$A_1^o = A_1^1 \cdot \Delta A + A_1^2 \cdot \Delta B + A_1^3 \cdot \Delta\alpha + A_1^4 \cdot \Delta\beta \quad (36a)$$

$$A_2^o = A_2^1 \cdot \Delta A + A_2^2 \cdot \Delta B + A_2^3 \cdot \Delta\alpha + A_2^4 \cdot \Delta\beta \quad (36b)$$

.

.

$$A_n^o = A_n^1 \cdot \Delta A + A_n^2 \cdot \Delta B + A_n^3 \cdot \Delta\alpha + A_n^4 \cdot \Delta\beta \quad (36c)$$

Minimizing the sum of squared deviations:

$$\Sigma[c_{fi} - (A \cdot e^{-\alpha t}\,i + B \cdot e^{-\beta t}\,i)] = \Sigma\, r_i = \min \quad (37)$$

leads to equations for ΔA, ΔB, $\Delta\alpha$ and $\Delta\beta$ (Krüger-Thiemer 1964). The corrections are added to the start parameters A^o, B^o, α^o, β^o and then the calculation is started again, using the corrected values as new start values until the correction satisfies the condition:

$$e = |\Delta A/A^o| + |\Delta B/B^o| + |\Delta\alpha/\alpha^o| + |\Delta\beta/\beta^o| \leq 10^{-6} \quad (38)$$

The values of A, B, α and β are then regarded as optimal estimators. If steady state conditions are not reached, thus A, B, α, and β shall be estimated from the uptake and the clearance phase, two different equations have to be used for one regression analysis, i.e. for the uptake:

$$c_f = A \cdot (1 - e^{-\alpha \cdot t}) + B \cdot (1 - e^{-\beta \cdot t}) \quad (15)$$

and:

$$c_f = A \cdot [e^{-\alpha \cdot (t-t^*)} - e^{-\alpha \cdot t}] + B \cdot [e^{-\beta \cdot (t-t^*)} - e^{-\beta \cdot t}] \quad (17)$$

for the clearance phase.

4.4 Examples and Discussion

4.4.1 Rate Constants and Half Life

In first order kinetics, rate constants are connected to half lifes by the general relationship:

$$t_{1/2} = \frac{\ln 2}{k} \tag{39}$$

This holds for the two- and the three-compartment model.

4.4.2 Comparison of Results: The Two-Compartment Compared to the Three Compartment Model

The results for the calculation of bioaccumulation factors from kinetic data using the two- and the three-compartment model may be demonstrated by two examples. The first example is the bioaccumulation of 3,4-dichloroaniline in fish (data of Nagel 1988) that leads to the results shown in Table 4.2 (see also Figure 4.5):

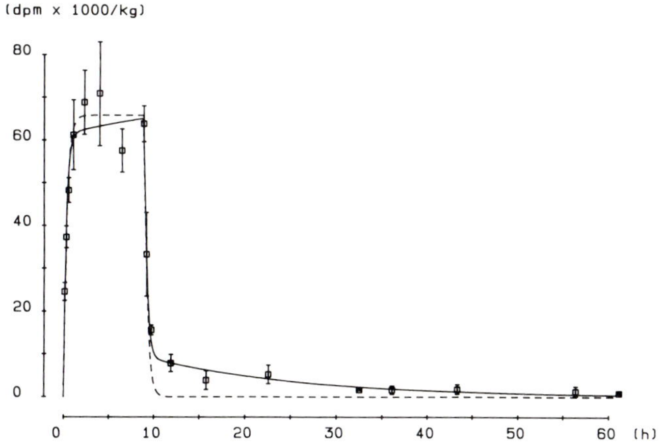

Figure 4.5 Uptake and Clearance of 3,4-Dichloroaniline in Fish

- - - : Two-Compartment Model,

—— : Three-Compartment Model

Table 4.2 BCF of 3,4-Dichloroaniline Calculated with Different Models

Two-Compartment Model	Three-Compartment Model (clearance only)	Three-Compartment Model (uptake & clearance)
30.8 (±4.7)	30.0 (±4.8)	39.3 (±25.3)

value of Nagel (1988): 30.2 (obtained from clearance data)

In this case the experiment was performed reaching steady state conditions. In the second example steady state conditions were not reached, it is the bioaccumulation of PCB 194 by mussels (data of Weigelt 1984). The results are compiled in Table 4.3 (see also Figure 4.6):

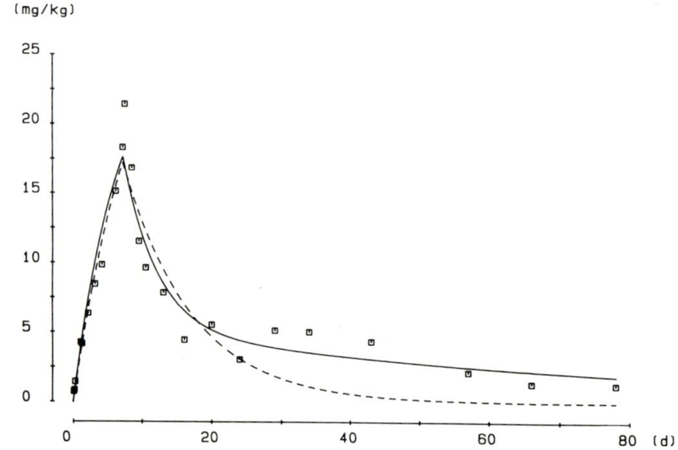

Figure 4.6 Uptake and Clearance of PCB 194 in Mussels
- - - : Two-Compartment Model,
—— : Three-Compartment Model

Table 4.3 BCF of PCB 194 Calculated with Different Methods

Two-Compartment Model	Three-Compartment Model (clearance only)	Three-Compartment Model (uptake & clearance)
48 100 (8740)	30 000 (9320)*	100 000 (60 000)

*clearance did not start under steady state conditions

Comparing the two- and the three-compartment model to describe the uptake and clearance of chemicals in a fish/water system one might summarize that both models lead to similar bioaccumulation factors, if the clearance phase starts under steady state conditions. The two models might lead to different results, if the clearance phase starts under non-equilibrium conditions.

The advantage of the two-compartment model is that less measurements are necessary to estimate the kinetic parameters (from which the BCF is calculated). On the other hand the three-compartment model leads to a more detailed description especially of the clearance phase. As shown in Figure 6 the two-compartment model might lead to a clearance rate constant, that simulates a quicker depuration of PCB 194 from the mussels than really measured. If models are applied, that are to simple, clearance of chemicals from organims might be overestimated, as chemicals might for a longer time be retained by organisms than assmued. This is especially dangerous regarding environmental conditions.

4.5 References

Blau, G.E., Neely, W.B. and D.R. Branson (1975). *AJChE Journal*, 21, 854-861.

Butte, W. and J.K. Blum (1984). *Chemosphere*, 13, 151-160.

Könemann, H. and K. van Leeuwen (1980). *Chemosphere*, 8, 3-19.

Knorre, W.A. (1981). Pharmakokinetik, Vieweg, Braunschweig.

Krüger-Thiemer, E. (1964). *Arzneimittelforsch.*, 14, 1332-1334.

Montgomery D.C. and E.A. Peck (1982). Introduction to Linear Regression Analysis, Wiley & Sons, New York.

Nagel, R. (1988). Umweltchemikalien und Fische - Beiträge zu einer Bewertung, Habilitationsschrift, Mainz.

Organisation of Economic Co-Operation and Development (1981). OECD Guidelines for Testing Chemicals: Bioaccumulation, OECD, Paris

Weigelt, V. (1984). Kapillargaschromatographische Untersuchungen polychlorierter Biphenyle (PCB) in marinen Organismen, Dissertation, Technische Universität Berlin.

4.6 Glossary

α	= rate constant (3-compartment model) [1/time]
A	= constant (3-compartment model) [mass/mass]
ß	= rate constant (3-compartment model) [1/time]
B	= constant (3-compartment model) [mass/mass]
BCF	= bioaccumulation factor [mass/volume]
BCF_n	= bioconcentration factor measured after n (5, 28, 56) days
BCF_∞	= bioconcentration factor measured under equilibrium condit.
c_f	= concentration of a substance in fish [mass/mass]
$c_{f\infty}$	= concentration in fish at steady state [mass/mass]
c_w	= concentration of a substance in water [mass/volume]
f_w	= fish weight [mass]
k_1	= uptake rate constant: water -> fish (2-compartment model) [1/time]
k_{12}	= uptake rate constant: water -> fish-compartment 2 (3-compartment model) [1/time]
k_2	= clearance rate constant: fish -> water (2-compartment model) [1/time]
k_{21}	= clearance rate constant: fish-compartment 2 - > water (3-compartment model) [1/time]
k_{23}	= rate constant: fish compartment 2 -> fish compartment 3 (3-compartment model) [1/time]
k_{32}	= rate constant: fish compartment 3 -> fish compartment 2 (3-compartment model) [1/time]
m_2	= mass of substance in fish compartment 2 [mass]
m_3	= mass of substance in fish compartment 3 [mass]
t	= time [time]
te	= time (if time scale starts with clearance) [time]
t*	= time for the beginning of the clearance phase [time]

QSARs of Bioconcentration: Validity Assessment of log P_{OW}/log BCF Correlations

Monika Nendza

5.1 Introduction

Chemical legislation, except Japan, does not request any experimental determination of bioconcentration potential in the basic tier, but relies on extrapolations from the compounds' physico-chemical properties, namely the logarithm of the 1-octanol/water partition coefficient (log P_{ow}). The simplistic evaluation procedure assumes no substantial bioconcentration for compounds having log P_{ow} < 2.7 - 3 (bioconcentration factor BCF < 100). Chemicals ranging in log P_{ow} between 2.7 - 3 and 6 are classified highly accumulating, eventually resulting in the demand for testing. Superlipophilic compounds characterized by log P_{ow} > 6 and molecular weight > 500 are regarded modestly accumulating. Basically, this assessment scheme represents the application of quantitative structure-activity relationships (QSARs) for legislative purposes. The scope of this study is to investigate the reliability and validity of the assumed log P_{ow}/log BCF correlations.

According to bioconcentration being defined as equilibrium partitioning between organisms and the surrounding medium, e.g. fish/water partitioning at steady state, modeling efforts have been based on analogous partitioning processes. Interphase distribution for application in QSAR studies is characterized by the 1-octanol/water partition coefficient P_{OW}. Since the lipid tissue of the fish is the principal site for bioaccumulation and 1-octanol is often a satisfactory surrogate for lipids, linear correlations are usually observed between log BCF and log P_{ow}. This corresponds to the underlying assumption of exchange between the water phase and the organic phases (e.g. fish) to be governed by diffusion processes. The Collander equation describes the assumed relationship between partitioning in different systems:

$$\log k_1 = a \log k_2 + b$$

with k_1 and k_2 being the corresponding partition coefficients (Collander 1951). The correlation, between e.g. BCF and P_{ow}, will be linear as long as the ratio of the respective activity coefficients remains constant. Several factors may cause deviations and apparent loss of linear correlation between log P_{ow} and log BCF, including experimental conditions, e.g. exposure time and concentration, metabolism, and bioavailability.

5.2 Structural Characteristics of Accumulating Compounds

Comparison of nonionic organic chemicals exhibiting substantial bioconcentration reveals several common characteristics (Table 5.1): The bioconcentration potential of a contaminant is directly related to its lipophilicity and inversely related to its water solubility, molecular charge and degree of ionization (Bysshe 1982, Connell 1988). Increasing size of the molecules, as expressed by the molecular weight, reduces the permeation through biological membranes to reach the site of potential accumulation, the upper limit being a molecular weight of about 500 (Umweltbundesamt 1990); Opperhuizen et al. (1985) reported a loss in membrane permeability with molecules having widths > 9.5 Å. Bioconcentration processes require a considerable period of time. Only upon continuous exposure, due to either persistence or continuous release, chemicals can reach the steady state. The stability and resistance to degradation is also reflected in soil persistence DT_{50} > 100 d.

Table 5.1 Structural characteristics of accumulating compounds

characteristic	bioconcentration is enhanced by:
lipophilicity	increasing log P_{OW} 0 - 6
	decline with P_{OW} > 6
water solubility	low solubility in aqueous phases
molecular charge	low degree of ionization
molecular size	molecular weight < 500
	molecular diameter < 9.5 Å
stability	low degree of transformation

5.3 Descriptors of Partitioning Processes

Hydrophobic chemicals are most likely to accumulate. Lipophilicity expressed by the 1-octanol/water partition coefficient under steady state conditions is the most prominent quantification of distribution behaviour. Alternative lipophilicity quantifications to the classical shake flask method are chromatographic procedures having the advantage that the dynamic transport in the column may be more similar to transport processes in the biophase. The slow-stirring method allows the experimental determination of high lipophilicity compounds (DeBruijn et al. 1989). Differences in the free energy of partitioning in different systems have been evaluated with respect to the individual contributions of solvation energy parameters (electrostatic, cavitation, hydrogen bonding terms) (Kamlet et al. 1988, Hawker 1990) and alternative solvent systems (Leahy et al. 1989). The availability of computerized log P_{ow} calculation (MedChem 1989) resulted in extensive use of this parameter. In general, the estimates are reliable, but for high log P_{ow} chemicals the results are subject to variant deviations.

Experimental and computational log P_{ow} determination is impeded for surface active compounds, organometallic (chelating) compounds, partly or fully dissociated acids and bases, chemicals of extremely high or low lipophilicity, mixtures, and impure compounds. The evident variability in parameterization of lipophilicity by e.g. log P_{ow} for description of BCFs (Schüürmann and Klein 1988) should be recognized, but it is only a minor factor introducing uncertainty as compared to the substantial problems arising from the quantification of the biological endpoint.

5.4 Established log P_{ow}/log BCF Correlations

Numerous QSARs estimating bioconcentration based on lipophilicity have been published. In general, an increase in BCF is associated with an increase in log P_{ow}. Examples of linear relationships are given in Table 5.2. The regression functions (eq. 1 - 6) can be discriminated with respect to their slopes, the regression coefficients are approx. 1.0 or 0.6 respectively. The unity slope reflects a strict dependence of BCF on partitioning solely, as it is also found in QSAR studies on the aquatic toxicity of nonpolar nonreactive toxicants. The lower slopes

Table 5.2 Linear log P_{ow}/log BCF_{FISH} Correlations:

$$\log BCF = a * \log P_{OW} + b$$

a	b	r	n	eq.	
0.54	0.12	0.95	8	(1)	Neely et al. (1974)
0.94	-1.95	0.87	26	(2)	Kenaga and Goring (1980)
0.79	-0.40	0.93	122	(3)	Veith and Kosian (1983)
1.00	-1.32	0.97	44 (?)	(4)	Mackay (1982)
1.02	-0.63	0.99	11	(5)	Oliver and Niimi (1983)
0.89	0.61	0.95	18	(6)	Chiou (1985)

a: regression coefficient, b: intercept, r: correlation coefficient, n: number of compounds analyzed of approx. 0.6 may be ascribed to the decreasing distribution rates of increasingly lipophilic chemicals as well as their limited solubility in the outer aqueous phase surrounding the organism and hindered transport over membrane barriers. The differences in the functions' intercepts, varying by more than two orders of magnitude between -2 and 0.6, have been attributed to the physiological differences of the tested fish, e.g. varying lipid content, and to the various classes of chemicals under study. Organic chemicals' BCFs for molluscs, mussels, and other aquatic invertebrates, microorganisms and the sediment to water system have also been related to log P_{ow} (Table 5.3).

Table 5.3 Linear log P_{ow}/log BCF Correlations:

$$\log BCF = a * \log P_{OW} + b$$

organism/ compartment	a	b	r	n	eq.	
mussel	0.86	-0.81	0.96	16	(7)	Geyer (1982)
oyster	0.49	1.03	0.62	14	(8)	Ogata et al. (1984)
molluscs	0.84	-1.23	0.83	34	(9)	Hawker et al. (1986)
microorg.	0.91	-0.36	0.98	14	(10)	Baughman et al. (1981)
sediment	1.00	-0.21	1.00	10	(11)	Karickhoff et al. (1979)

a: regression coefficient, b: intercept, r: correlation coefficient, n: number of compounds analyzed

A loss of linear correlation has been observed for the range of high lipophilicity. To account for the reduced bioconcentration potential of superlipophilic compounds, non-linear QSAR models have been derived (Table 5.4). A parabolic relationship has been developed by Könemann and van Leeuwen (1980) for chlorobenzenes. Connell and Hawker (1988) derived a polynominal log P_{ow} dependent function to describe BCFs of chlorinated hydrocarbons with a maximum bioconcentration for compounds with log P_{ow} 6.7. The section of this curve between log P_{ow} 3 and 6 is approximately linear and coincides with the linear equations (eq. 3, 4, table 2). Spacie and Hamelink (1982) proposed a sigmoid model, without stating the corresponding function, to account for the fact that the linear correlations also break down for very hydrophilic compounds.

5.5 QSARs Variability due to Variance in Underlying Experimental Data

A major limitation to the derivation of reliable QSARs and eventually the prediction of

BCF values is posed by the substantial variability in measured BCF values. The data may range over several orders of magnitude, depending on e.g. the compounds' purity and exposure concentration, loss of chemical by evaporation, test species and protocol. Especially for very lipophilic chemicals the determination of the concentration in the aqueous solution is difficult and therefore, differences between actual and measured exposure may result in further bias of the obtained BCF values. Also, considerable variability in log P_{OW} in the high lipophilicity range introduces further uncertainty.

Table 5.4 Non-linear log P_{OW}/log BCF_{FISH} Correlations:

eq. (12) Könemann and van Leeuwen (1980)
$\log BCF = 3.41 \log P_{OW} - 0.26 (\log P_{OW})^2 - 5.51$
r = n.a. n = 6

eq. (13) Connell and Hawker (1988)
$\log BCF = 0.0069(\log P_{OW})^4 - 0.185(\log P_{OW})^3 + 1.55(\log P_{OW})^2 - 4.18 \log P_{OW} + 4.79$
r = n.a. n = 45-46 (?)

eq. (14) this study
$\log BCF = 0.99 \log P_{OW} - 1.47 \log (4.97 * 10^{-8} P_{OW} + 1) + 0.0135$

r: correlation coefficient, n: number of compounds analyzed, n.a.: value not given

Evidently, the preciseness of QSAR predictions can not exceed that of the underlying data. QSAR modeling is limited by the fact that BCF does not represent a well defined, consistent endpoint. BCF has to be recognized a species specific parameter, even correction for the lipid content of the organisms is not sufficient for normalization. The existing protocols for BCF determination hardly result in the homogeneous data suited for comparative analysis.

5.5.1 QSARs Variability due to Non-Equilibrium Partitioning during Experiments

The time needed to reach equilibrium between the water and the fish increases with increasing contaminants' lipophilicity. For organisms such as crustaceans with a short life-span relative to the biological half lives of the chemicals, steady state concentration may not be achieved. Chemicals which do not reach equilibrium during the experiment are characterized by high partition coefficients (log P_{ow} > 6). Their experimental BCFs are well below those predicted from log P_{ow}/log BCF relationships. Oliver (1984) reported for QSARs considering only compounds that reached equilibrium the slope of the corresponding regression function to be approx. 1.0 (unity). If highly lipophilic compounds are included in the data set, the exposure time effect causes a levelling of the log P_{ow}/log BCF functions' slope.

5.5.2 QSARs Variability due to Variant Bioavailability

Chemicals having log P_{ow} > 6 are mostly of very low aqueous solubility (< 10 μg/l), and a major fraction will be adsorbed by suspended particles or colloidal organic matter in the water phase (Bruggeman et al. 1984, Schrap and Opperhuizen 1990). The presently applied water analyses rarely distinguish between the bioavailable and non-bioavailable chemical in the water resulting in a miscalculation of the bioconcentration potential. Empirical uptake rate constants reveal a similar course, the activity in the water is much lower than assumed because the xenobiotic is adsorbed to dissolved organics.

5.6 Reliability of QSAR Estimated BCF Values

The confidence in the established QSARs for predictive purposes suffers from the uncertainties of the input data (BCF and log P_{ow}) used for derivation of the functions. For 132 randomly selected compounds, for which BCFs are published (Bruggeman et al. 1984, Oliver 1984, Butte et al. 1987, Deneer et al. 1987, Gobas et al. 1987, Sabljic 1987, Gobas and Schrap 1990, Hawker 1990, Smith et al. 1990), log P_{ow} values were calculated (MedChem 1989). The trend of BCFs increasing with log P_{ow} can be easily recognized, revealing also the substantial variation in BCFs for compounds of approximately identical lipophilicity by at least two orders of magnitude (Figure 5.1). The respective data points are contrasted with QSAR functions (Figures 5.2 and 5.3) revealing the significant deviations (a) among the curves, and (b) between the predicted and measured data. If the log P_{ow}/log BCF relation is restricted to a subset of compounds for which BCFs are normalized for lipid content of the fish (Figure 5.4), the situation does not change significantly. Inspection of the data reveals clustering with respect to chemical classes.

Comparison of the QSARs relative to the experimental data reveals that most functions result in estimated "average" BCFs, which may be exceeded by measured values by one order of magnitude.

For the entire data set of BCF values, it is evident that a non-linear function can be constructed based on log P_{ow} which describes the highest BCF associated with a given lipophilicity. If then discrepancies between measured and calculated values occur, the measured BCFs

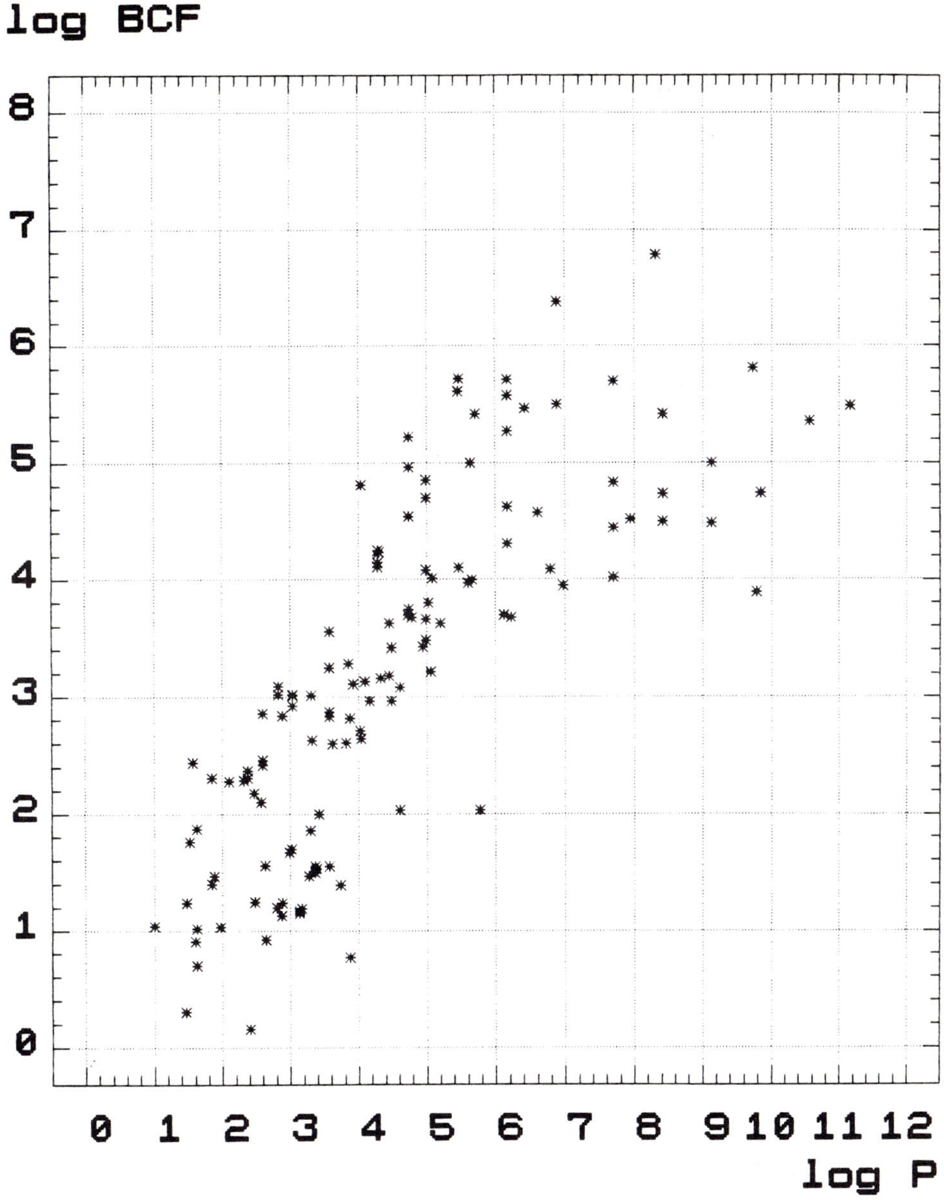

Figure 5.1 Relationship between 132 randomly selected experimental log BCF data and calculated log P_{ow}. (BCF data taken from: Bruggeman et al. 1984, Oliver 1984, Butte et al. 1987, Deneer et al. 1987, Sabljic 1987, Gobas et al. 1990, Hawker 1990, Smith et al. 1990)

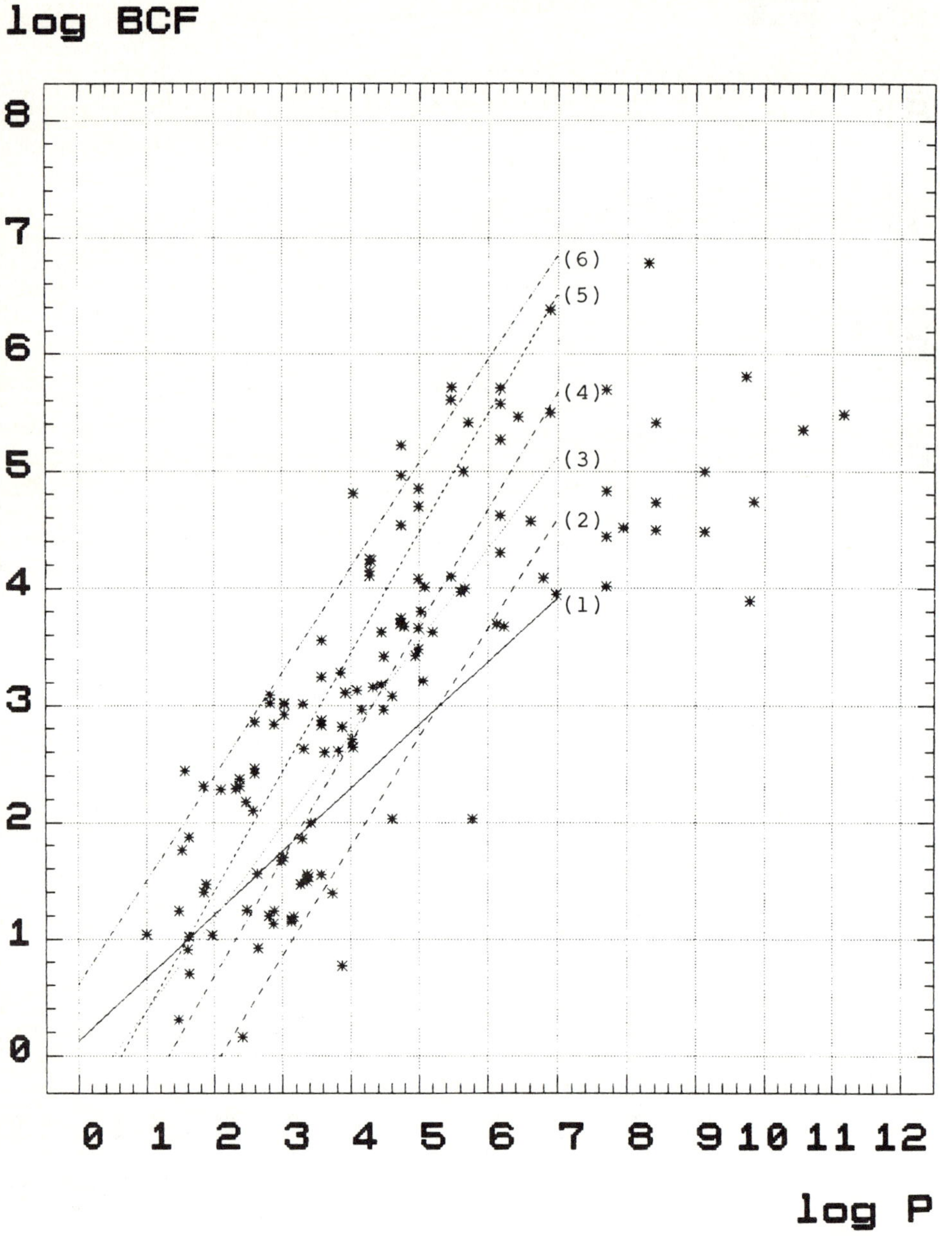

Figure 5.2 Comparison of experimental BCF data for 132 randomly selected compounds with linear log P_{ow}/log BCF correlations. For identification of QSARs see Table 5.2. (BCF data references see Figure 5.1)

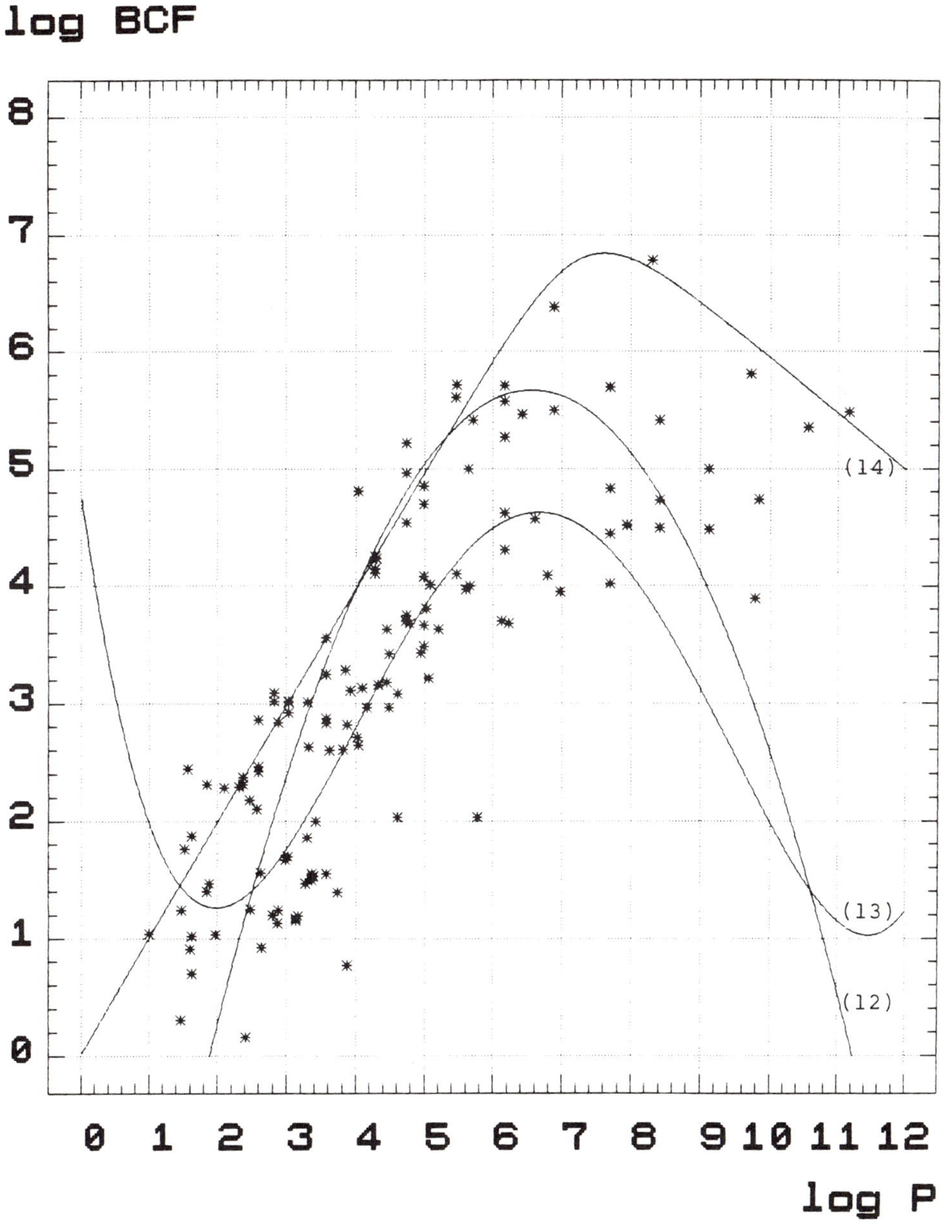

Figure 5.3 Comparison of experimental BCF data for 132 randomly selected compounds with non-linear log P_{ow}/ log BCF correlations. For identification of QSARs see Table 5.4. BCF data references see Figure 5.1)

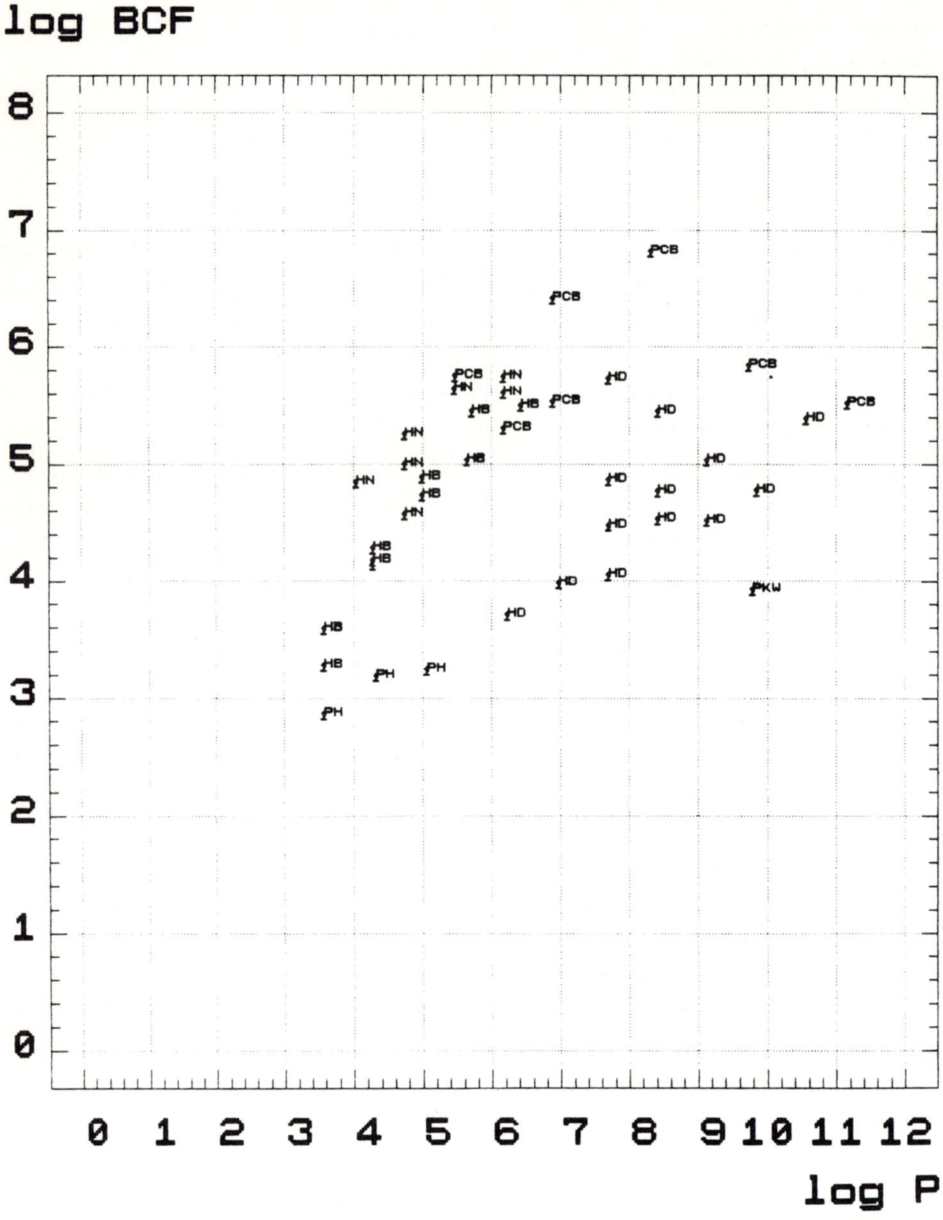

Figure 5.4 Relationship of BCFs normalized for fish lipid content on log P_{ow} discriminating classes of compounds. HB: halogenated benzenes, HD: halogenated dibenzo-p-dioxins, HN: halogenated naphthalenes, PKW: polycyclic hydrocarbons, PH: phenols, PCB: polychlorinated biphenyls. (BCF data references see Figure 5.1)

are lower than calculated. Thus, this function will result in "worst case" estimates of bioconcentration and can be expressed as:

$$\log BCF = 0.99 \log P_{ow} - 1.47 \log (4.97 * 10^{-8} P_{ow} + 1) + 0.0135 \qquad \text{(eq.14)}$$

Standard statistical parameters are not meaningful with this equation, and hence are not given, as it does not represent a regression on BCF data, but a description of the hypothetical "worst case" function (Table 5.4). This formalization resumes a bilinear curve with a linearly increasing part between log P_{ow} 0 and 6. The empirically postulated coincidence of log P_{ow} and log BCF in this lipophilicity range is reflected by the approx. unity slope (0.99) for the 1st order log P_{ow} term and the intercept of about 0. Maximum log BCF values of approx. 7 ensue from log P_{ow} 7, which is in coincidence with previously reported results (Könemann and van Leeuwen 1980, Connell and Hawker 1988). More lipophilic compounds are expected to be less accumulating corresponding to the negative slope derived for the second log P_{ow} term of equation 14.

Comparison of measured log BCF values with estimates obtained from equations 1 - 6 and equations 12 - 14 (Table 5.5) reveals considerable differences in the goodness of fit of the predictions. The linear functions (eq. 1 - 6) were analyzed separately as they are not applicable in the high lipophilicity range. These QSARs result in the same relative ranking of accumulation potential due to the consideration of the same principal property (log P_{ow}). Inspection of the respective residuals (log $BCF_{obs.}$- log $BCF_{calc.}$) reveals some overestimates but, more severely, also much underrating of BCFs. An exception is equation 6 by providing no significant underestimates, even though this QSAR is biased due to the consideration of only few ($n = 13$) compounds representing a very narrow log P_{ow} range. The non-linear functions (eq. 12 - 14) have a broader application range extending to the domain of superlipophilic compounds. The QSARs equation 12 and equation 13 also do not represent "worst case" models, resulting in estimates that may be up to 2 orders of magnitude below experimental BCFs. Only equation 14 results in predominantly negative residuals, hence reflecting the "worst case".

5.6.1 Parameter Range Limitations

To apply any existing QSAR model, it is essential to assess its range of application with respect to the structural and physico-chemical parameter space covered. The chemical, for which BCF is to be predicted, should be similar to the structures used for deriving the model and its log P_{ow} should be within the parameter range covered. The variable domain spanned by a QSAR should be homogeneously represented by a substantial number of compounds and

Table 5.5 Comparison of experimental BCF data for 132 randomly selected compounds with predictions obtained from log P_{ow}/log BCF correlations. For identification of QSARs see Tables 5.2 and 5.4. (BCF data taken from: Bruggeman et al. 1984, Oliver 1984, Butte et al. 1987, Deneer et al. 1987, Sabljic 1987, Gobas et al. 1990, Hawker 1990, Smith et al. 1990)

Compound	log Pow calc.	log BCF exp.	log BCF calc. eq(1)	log BCF calc. eq(2)	log BCF calc. eq(3)	log BCF calc. eq(4)	log BCF calc. eq(5)	log BCF calc. eq(6)	log BCF calc. eq(12)	log BCF calc. eq(13)	log BCF calc. eq(14)
2,3,4-Trichloroanisole	4.16	2.96	2.38	1.94	2.89	2.84	3.62	4.32	4.11	2.97	4.13
2,4-Dichlorodiphenylether	5.60	3.97	3.16	3.29	4.02	4.28	5.09	5.61	5.31	4.29	5.53
2,4,2',4'-Tetrachlorodiphenyloxide	6.79	4.09	3.80	4.40	4.96	5.47	6.31	6.67	5.48	4.62	6.34
Diethylphthalate	2.57	2.10	1.52	0.46	1.63	1.25	1.99	2.90	1.51	1.45	2.56
Dimethylphthalate	1.52	1.76	0.95	-.53	0.80	0.20	0.92	1.96	-0.94	1.40	1.52
1,2-Dimethylbenzene	3.14	1.15	1.83	0.99	2.08	1.82	2.58	3.41	2.59	1.89	3.12
1,3-Dimethylbenzene	3.14	1.17	1.83	0.99	2.08	1.82	2.58	3.41	2.59	1.89	3.12
1,4-Dimethylbenzene	3.14	1.17	1.83	0.99	2.08	1.82	2.58	3.41	2.59	1.89	3.12
Ethylbenzene	3.17	1.19	1.84	1.02	2.10	1.85	2.61	3.44	2.65	1.92	3.15
Isopropylbenzene	3.57	1.55	2.06	1.39	2.42	2.25	3.02	3.80	3.30	2.33	3.55
Styrene	2.87	1.13	1.68	0.74	1.87	1.55	2.30	3.17	2.10	1.66	2.85
Toluene	2.64	0.92	1.55	0.52	1.69	1.32	2.07	2.96	1.65	1.49	2.63
m-Methylstyrene	3.37	1.55	1.95	1.20	2.26	2.05	2.81	3.62	2.98	2.12	3.35
p-Methylstyrene	3.37	1.50	1.95	1.20	2.26	2.05	2.81	3.62	2.98	2.12	3.35
β-Methylstyrene	3.39	1.53	1.96	1.22	2.28	2.07	2.83	3.63	3.02	2.14	3.37
α-Methylstyrene	3.27	1.47	1.90	1.11	2.18	1.95	2.71	3.53	2.82	2.02	3.25
1,2-Dichlorobenzene	3.57	3.56	2.06	1.39	2.42	2.25	3.02	3.80	3.30	2.33	3.55
1,2,3-Trichlorobenzene	4.28	4.11	2.44	2.05	2.98	2.96	3.74	4.43	4.25	3.10	4.25
1,2,3,4-Tetrachlorobenzene	4.99	4.08	2.83	2.72	3.54	3.67	4.47	5.06	4.93	3.82	4.95
1,2,3,5-Tetrachlorobenzene	4.99	4.85	2.83	2.72	3.54	3.67	4.47	5.06	4.93	3.82	4.95
1,2,4-Tribromobenzene	4.45	3.63	2.54	2.21	3.12	3.13	3.92	4.58	4.44	3.29	4.42
1,2,4-Trichlorobenzene	4.28	4.25	2.44	2.05	2.98	2.96	3.74	4.43	4.25	3.10	4.25
1,2,4,5-Tetrabromobenzene	5.03	3.80	2.85	2.76	3.57	3.71	4.51	5.10	4.96	3.85	4.99
1,2,4,5-Tetrachlorobenzene	4.99	4.70	2.83	2.72	3.54	3.67	4.47	5.06	4.93	3.82	4.95
1,3-Dibromobenzene	3.87	2.82	2.22	1.67	2.66	2.55	3.32	4.06	3.73	2.65	3.84
1,3-Dichlorobenzene	3.57	2.87	2.06	1.39	2.42	2.25	3.02	3.80	3.30	2.33	3.55
1,3,5-Tribromobenzene	4.73	3.70	2.69	2.48	3.34	3.41	4.20	4.83	4.71	3.57	4.69
1,3,5-Trichlorobenzene	4.28	4.14	2.44	2.05	2.98	2.96	3.74	4.43	4.25	3.10	4.25
1,4-Dichlorobenzene	3.57	3.25	2.06	1.39	2.42	2.25	3.02	3.80	3.30	2.33	3.55
2,4,5-Trichlorotoluene	4.78	3.68	2.71	2.52	3.38	3.46	4.25	4.88	4.76	3.62	4.74
3,4-Dichlorobenzotrifluoride	4.45	3.18	2.54	2.21	3.12	3.13	3.92	4.58	4.44	3.29	4.42
Hexabromobenzene	5.64	5.00	3.18	3.33	4.06	4.32	5.13	5.64	5.33	4.31	5.57
Hexachlorobenzene	6.42	5.46	3.60	4.06	4.67	5.10	5.93	6.34	5.50	4.61	6.19
Octachlorostyrene	7.94	4.52	--	--	--	--	--	--	4.93	4.14	5.41
Pentachlorobenzene	5.71	5.41	3.22	3.39	4.11	4.39	5.20	5.71	5.36	4.35	5.63
1,2,3,4-Tetrachlorodibenzo-p-dioxin	7.70	4.02	--	--	--	--	--	--	5.10	4.30	5.80
1,2,3,4,6,7,8-Heptachlorodibenzo-p-dioxin	9.85	4.74	--	--	--	--	--	--	2.47	2.15	1.14
1,2,3,4,7-Pentachlorodibenzo-p-dioxin	8.41	4.50	--	--	--	--	--	--	4.50	3.74	4.48
1,2,3,4,7,8-Hexachlorodibenzo-p-dioxin	9.13	5.00	--	--	--	--	--	--	3.62	2.98	2.85
1,2,3,7-Tetrachlorodibenzo-p-dioxin	7.70	4.44	--	--	--	--	--	--	5.10	4.30	5.80
1,2,3,7,8-Pentachlorodibenzo-p-dioxin	8.41	5.41	--	--	--	--	--	--	4.50	3.74	4.48
1,2,3,7,8,9-Hexachlorodibenzo-p-dioxin	9.13	4.48	--	--	--	--	--	--	3.62	2.98	2.85
1,2,3,7,9-Pentachlorodibenzo-p-dioxin	8.41	4.73	--	--	--	--	--	--	4.50	3.74	4.48
1,2,4-Trichlorodibenzo-p-dioxin	6.97	3.95	3.90	4.57	5.11	5.65	6.49	6.83	5.44	4.60	6.35
1,3,6,8-Tetrachlorodibenzo-p-dioxin	7.70	4.83	--	--	--	--	--	--	5.10	4.30	5.80
2,3,7,8-Tetrachlorodibenzo-p-dioxin	7.70	5.70	--	--	--	--	--	--	5.10	4.30	5.80
2,7-Dichlorodibenzo-p-dioxin	6.23	3.68	3.50	3.88	4.52	4.91	5.74	6.17	5.49	4.57	6.06
Octachlorodibenzo-p-dioxin	10.56	5.35	--	--	--	--	--	--	1.07	1.45	-0.56

Table 5.5 continued

Compound	log Pow calc.	log BCF exp.	log BCF calc. eq(1)	log BCF calc. eq(2)	log BCF calc. eq(3)	log BCF calc. eq(4)	log BCF calc. eq(5)	log BCF calc. eq(6)	log BCF calc. eq(12)	log BCF calc. eq(13)	log BCF calc. eq(14)
2,4-Dichloroethane	1.46	0.30	0.92	-.58	0.75	0.14	0.86	1.91	-1.10	1.45	1.46
Aldrin	5.09	4.01	2.88	2.81	3.62	3.77	4.57	5.15	5.01	3.91	5.04
Bis(2-chloroethyl)ether	1.00	1.04	0.67	-1.0	0.39	-.32	0.39	1.50	-2.37	1.98	1.00
Carbontetrachloride	2.88	1.24	1.68	0.75	1.88	1.56	2.31	3.18	2.12	1.66	2.86
Hexachlorobutadiene	4.30	4.23	2.45	2.07	3.00	2.98	3.76	4.45	4.27	3.13	4.27
Hexachloroethan	4.61	3.08	2.62	2.36	3.24	3.29	4.08	4.72	4.60	3.45	4.57
Tetrachloroethene	3.02	1.70	1.76	0.88	1.99	1.70	2.45	3.30	2.38	1.78	3.00
Trichloroethene	2.80	1.20	1.64	0.67	1.81	1.48	2.23	3.11	1.97	1.60	2.79
o,p'-DDT	6.61	4.57	3.71	4.23	4.82	5.29	6.12	6.51	5.50	4.63	6.29
1,2,3,4-Tetrachloronaphthalene	6.17	5.71	3.47	3.82	4.47	4.85	5.67	6.12	5.48	4.55	6.02
1,3,5,7-Tetrachloronaphthalene	6.17	5.71	3.47	3.82	4.47	4.85	5.67	6.12	5.48	4.55	6.02
1,3,5,8-Tetrachloronaphthalene	6.17	5.57	3.47	3.82	4.47	4.85	5.67	6.12	5.48	4.55	6.02
1,3,7-Trichloronaphthalene	5.45	5.61	3.08	3.15	3.91	4.13	4.94	5.47	5.24	4.19	5.39
1,4-Dichloronaphthalene	4.74	4.54	2.69	2.48	3.34	3.42	4.21	4.84	4.72	3.58	4.70
1,4-Dichloronaphthalene	4.74	3.75	2.69	2.48	3.34	3.42	4.21	4.84	4.72	3.58	4.70
1,8-Dichloronaphthalene	4.74	4.96	2.69	2.48	3.34	3.42	4.21	4.84	4.72	3.58	4.70
2-Chloronaphthalene	4.03	4.81	2.31	1.82	2.78	2.71	3.49	4.21	3.95	2.83	4.00
2,3-Dichloronaphthalene	4.74	5.22	2.69	2.48	3.34	3.42	4.21	4.84	4.72	3.58	4.70
2,7-Dichloronaphthalene	4.74	5.22	2.69	2.48	3.34	3.42	4.21	4.84	4.72	3.58	4.70
1,2-Dinitrobenzene	1.63	1.02	1.01	-.42	0.89	0.31	1.03	2.06	-0.65	1.34	1.63
1,3-Dinitrobenzene	1.63	1.87	1.01	-.42	0.89	0.31	1.03	2.06	-0.65	1.34	1.63
1,4-Dinitrobenzene	1.63	0.70	1.01	-.42	0.89	0.31	1.03	2.06	-0.65	1.34	1.63
2-Chloronitrobenzene	2.32	2.29	1.38	0.22	1.43	1.00	1.74	2.68	0.98	1.32	2.31
2-Chloro-6-Nitrotoluene	2.82	3.09	1.65	0.69	1.83	1.50	2.25	3.13	2.01	1.62	2.81
2-Nitrotoluene	2.10	2.28	1.26	0.02	1.26	0.78	1.51	2.48	0.49	1.27	2.09
2,3-Dichloronitrobenzene	3.03	3.01	1.77	0.89	1.99	1.71	2.46	3.31	2.40	1.79	3.01
2,3-Dimethylnitrobenzene	2.60	2.86	1.53	0.48	1.65	1.28	2.03	2.93	1.57	1.46	2.59
2,4-Dichloronitrobenzene	3.03	3.02	1.77	0.89	1.99	1.71	2.46	3.31	2.40	1.79	3.01
2,4-Dinitrotoluene	1.85	2.31	1.13	-.22	1.06	0.53	1.26	2.26	-0.11	1.27	1.84
2,5-Dichloronitrobenzene	3.03	2.92	1.77	0.89	1.99	1.71	2.46	3.31	2.40	1.79	3.01
2,6-Dinitrotoluene	1.57	2.44	0.97	-.48	0.84	0.25	0.97	2.01	-0.81	1.37	1.57
3-Chloronitrobenzene	2.60	2.42	1.53	0.48	1.65	1.28	2.03	2.93	1.57	1.46	2.59
3-Nitrotoluene	2.38	2.31	1.41	0.28	1.48	1.06	1.80	2.73	1.11	1.35	2.37
3,4-Dimethylnitrobenzene	2.88	2.84	1.68	0.75	1.88	1.56	2.31	3.18	2.12	1.66	2.86
3,5-Dichloronitrobenzene	3.31	3.01	1.92	1.15	2.21	1.99	2.75	3.56	2.88	2.06	3.29
4-Chloronitrobenzene	2.60	2.46	1.53	0.48	1.65	1.28	2.03	2.93	1.57	1.46	2.59
4-Chloro-2-Nitrotoluene	2.82	3.02	1.65	0.69	1.83	1.50	2.25	3.13	2.01	1.62	2.81
4-Nitrotoluene	2.38	2.37	1.41	0.28	1.48	1.06	1.80	2.73	1.11	1.35	2.37
Nitrobenzene	1.88	1.47	1.14	-.19	1.09	0.56	1.29	2.29	-0.03	1.27	1.87
2,2',3,3',4,4',5,5'-Octachlorobiphenyl	9.73	5.81	--	--	--	--	--	--	2.68	2.29	1.43
2,2',4,4',5,5'-Hexachlorobiphenyl	8.31	6.78	--	--	--	--	--	--	4.60	3.83	4.70
2,2',5-Trichlorobiphenyl	6.17	4.30	3.47	3.82	4.47	4.85	5.67	6.12	5.48	4.55	6.02
2,2',5,5'-Tetrachlorobiphenyl	6.88	6.38	3.85	4.49	5.04	5.56	6.40	6.75	5.46	4.61	6.35
2,3',4',5-Tetrachlorobiphenyl	6.88	5.50	3.85	4.49	5.04	5.56	6.40	6.75	5.46	4.61	6.35
2,4,5-Trichlorobiphenyl	6.17	5.27	3.47	3.82	4.47	4.85	5.67	6.12	5.48	4.55	6.02
2,4',5-Trichlorobiphenyl	6.17	4.62	3.47	3.82	4.47	4.85	5.67	6.12	5.48	4.55	6.02
2,5-Dichlorobiphenyl	5.46	5.72	3.08	3.16	3.91	4.14	4.95	5.48	5.24	4.19	5.40
4,4'-Dichlorobiphenyl	5.46	4.10	3.08	3.16	3.91	4.14	4.95	5.48	5.24	4.19	5.40

Table 5.5 continued

Compound	log Pow calc.	log BCF exp.	log BCF calc. eq(1)	log BCF calc. eq(2)	log BCF calc. eq(3)	log BCF calc. eq(4)	log BCF calc. eq(5)	log BCF calc. eq(6)	log BCF calc. eq(12)	log BCF calc. eq(13)	log BCF calc. eq(14)
Biphenyl	4.03	2.64	2.31	1.82	2.78	2.71	3.49	4.21	3.95	2.83	4.00
Decachlorobiphenyl	11.16	5.48	--	--	--	--	--	--	-0.33	1.08	-1.99
2-Methylphenol	1.97	1.03	1.19	-.11	1.16	0.65	1.38	2.37	0.18	1.26	1.96
2-Methyl-4,6-Dinitrophenol (DNOC)	2.41	0.16	1.43	0.31	1.50	1.09	1.83	2.76	1.17	1.36	2.40
2-Sec-butyl-4,6-dinitrophenol (Dinoseb)	3.87	0.77	2.22	1.67	2.66	2.55	3.32	4.06	3.73	2.65	3.84
2-Tert-butyl-4,6-dinitrophenol(Dinoterb)	3.74	1.39	2.15	1.55	2.55	2.42	3.19	3.95	3.55	2.51	3.72
2,3,5,6-Tetrachlorophenol	4.32	3.16	2.47	2.09	3.01	3.00	3.78	4.46	4.30	3.15	4.29
2,4-Dimethylphenol	2.47	2.18	1.46	0.36	1.55	1.15	1.89	2.81	1.30	1.39	2.46
2,4,5-Trichlorophenol	3.85	3.28	2.21	1.65	2.64	2.53	3.30	4.05	3.71	2.63	3.82
2,4,6-Tribromophenol	4.02	2.71	2.30	1.81	2.78	2.70	3.48	4.20	3.93	2.82	3.99
2,4,6-Trichlorophenol	3.57	2.83	2.06	1.39	2.42	2.25	3.02	3.80	3.30	2.33	3.55
2,6-Dibromo-4-Cyanophenol (Bromoxynil)	2.99	1.67	1.74	0.85	1.96	1.67	2.42	3.28	2.33	1.76	2.97
3-Chlorophenol	2.48	1.25	1.47	0.37	1.56	1.16	1.90	2.82	1.32	1.40	2.47
4-Bromophenol	2.63	1.56	1.55	0.51	1.68	1.31	2.06	2.96	1.63	1.48	2.62
4-Cyanophenol	1.60	0.91	0.99	-.45	0.86	0.28	1.00	2.04	-0.73	1.36	1.60
4-Tert-butylphenol	3.30	1.86	1.91	1.14	2.21	1.98	2.74	3.55	2.87	2.05	3.28
Pentachlorophenol	5.06	3.21	2.87	2.78	3.60	3.74	4.54	5.13	4.99	3.88	5.01
Phenol	1.48	1.24	0.93	-.56	0.77	0.16	0.88	1.93	-1.04	1.43	1.48
3-Nitrophenol	1.85	1.40	1.13	-.22	1.06	0.53	1.26	2.26	-0.11	1.27	1.84
2-Chlorophenanthrene	5.20	3.63	2.94	2.92	3.71	3.88	4.68	5.25	5.09	4.00	5.15
2-Methylphenanthrene	4.99	3.48	2.83	2.72	3.54	3.67	4.47	5.06	4.93	3.82	4.95
9-Methylanthracene	4.99	3.66	2.83	2.72	3.54	3.67	4.47	5.06	4.93	3.82	4.95
Acenaphthene	3.62	2.60	2.09	1.44	2.46	2.30	3.07	3.84	3.38	2.38	3.60
Acridine	3.43	2.00	1.98	1.26	2.31	2.11	2.87	3.67	3.08	2.18	3.41
Anthracene	4.49	2.96	2.56	2.25	3.15	3.17	3.96	4.62	4.48	3.33	4.46
Benzo(a)pyrene	6.12	3.70	3.44	3.78	4.43	4.80	5.62	6.07	5.47	4.54	5.98
Benz(a)acridine	4.61	2.03	2.62	2.36	3.24	3.29	4.08	4.72	4.60	3.45	4.57
Benz(a)anthracene	5.66	4.00	3.19	3.35	4.07	4.34	5.15	5.66	5.34	4.32	5.58
Dibenzofuran	4.09	3.13	2.34	1.88	2.83	2.77	3.55	4.26	4.02	2.90	4.06
Dibenz(a,h)acridine	5.78	2.03	3.26	3.46	4.17	4.46	5.28	5.77	5.38	4.39	5.69
Fluorene	3.92	3.11	2.25	1.72	2.70	2.60	3.37	4.11	3.80	2.71	3.89
Octachlorodibenzofuran	9.79	3.89	--	--	--	--	--	--	2.58	2.22	1.28
Phenanthrene	4.49	3.42	2.56	2.25	3.15	3.17	3.96	4.62	4.48	3.33	4.46
Pyrene	4.95	3.43	2.81	2.68	3.51	3.63	4.43	5.03	4.90	3.78	4.91
2-Methylnaphthalene	3.82	2.61	2.19	1.62	2.62	2.50	3.27	4.02	3.66	2.60	3.79
Naphthalene	3.32	2.63	1.92	1.16	2.22	2.00	2.76	3.57	2.90	2.07	3.30
Residuals			85.36	146.23	33.00	43.82	-45.99	-130.90	-10.72	54.03	-76.81

can be depicted by scatter plots of the variables (figures 5 A - G) if the necessary data are given. The QSAR models equations 1 - 6 mostly consider compounds having log P_{ow} > 2 and < 6. The omission of superlipophilic chemicals is a prerequisite for deriving linear functions, which hence are not applicable for predicting BCFs if the compounds' log P_{ow} exceeds 6. The result of neglecting compounds in the low log P_{ow} range is comprehensively illustrated by the non-linear model equation 13, which results in higher BCF estimates for compounds having log P_{ow} 0 than for those having log P_{ow} 5 (the calculated log BCF is 4.8 and 3.8 respectively). This experience emphasizes again, that prediction may be hazardous when obtained by extrapolating outside the parameter space investigated.

5.6.2 Deviations due to Extreme Lipophilicity

Chemicals with log P_{ow} > 6 have lower BCFs than calculated from linear QSARs. For these compounds, BCFs no longer increase with respect to log P_{ow}. A maximum of log BCF approx. 6 - 7 for compounds with log P_{ow} 6 - 8 is observed, then a gradual decrease with further increase in log P_{ow} occurs.

Investigations of water to 1-octanol transfer for hydrophobic compounds revealed the rate constants to be essentially independent of log P_{ow} (Hawker and Connell 1989) indicating that aqueous phase diffusion is the controlling factor for these solutes. For an extended log P_{ow} range, a curvilinear relationship has to be expected between the logarithm of the water to 1-octanol transfer rate constants and log P_{ow}, as has been also found for water to lipid transfer or uptake to aquatic organisms. These qualitative similarities in mass transfer kinetics in abiotic and biotic partitioning systems suggest analogous control processes for lipid/water and 1-octanol/water systems, but at different solute P_{ow} values and with different magnitudes of rate constants.

Differences in the thermodynamic properties of lipid/water and 1-octanol/water partitioning processes, e.g. enthalpy changes, have been observed for different types of lipophilic chemicals indicating that no unique log P_{ow}/log BCF relationship can be assumed for all contaminants (Opperhuizen et al. 1988).

Comparing the phase properties of the fish lipids and 1-octanol towards organic chemicals reveals different structures of the lipid phases. The fish lipid consists primarily of biological membranes, in which the molecules are predominantly arranged in bilayers. The lipid phase has thus a distinct structure and restricted spatial dimensions. Since the 1-octanol phase is a bulk phase presumably with little or no structure, organic solutes may display different activity coefficients and partitioning behaviour in 1-octanol than in membranes. The loss of linear correlation between log P_{ow} and log BCF can then be at least partly ascribed to differences in

A

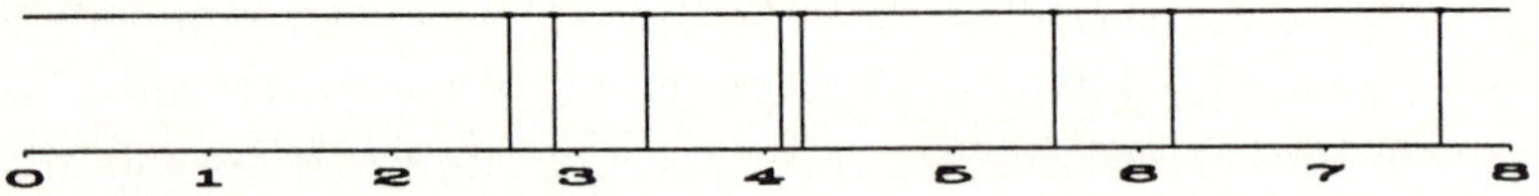

eq. 1 (Neely et al. 1974)
compound class: organochlorine and aromatic hydrocarbons
species: Salmo gairdneri
log P_{ow}: 2.64 - 7.62

B

eq. 3 (Veith and Kosian 1983)
compound class: 122 miscellaneous compounds
species: 13 species of freshwater and marine fish
log P_{ow}: 1.00 - 6.91

C

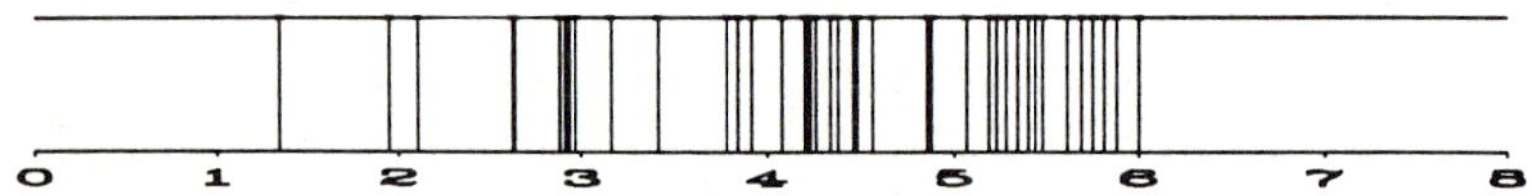

eq. 4 (Mackay 1982)
compound class: 44 (?) miscellaneous compounds
species: fish (no further specification)
log P_{ow}: 1.35 - 6.00

D

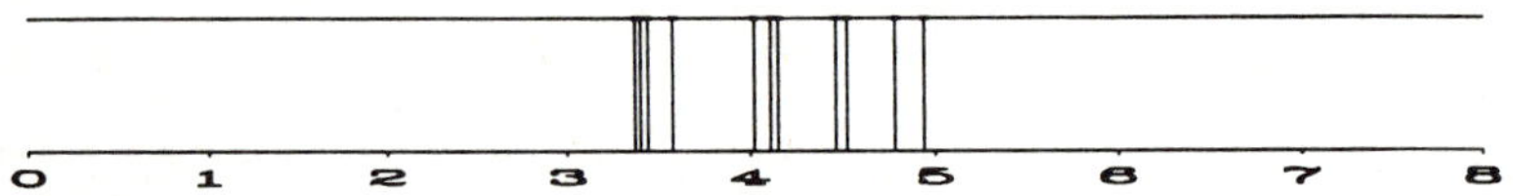

eq. 5 (Oliver 1984)
compound class: 11 chlorobenzenes
species: Salmo gairdneri
log P_{ow}: 3.37 - 4.94

Figure 5.5 Assessment of the log P_{ow} parameter covered by BCF_{FISH}QSARs

E

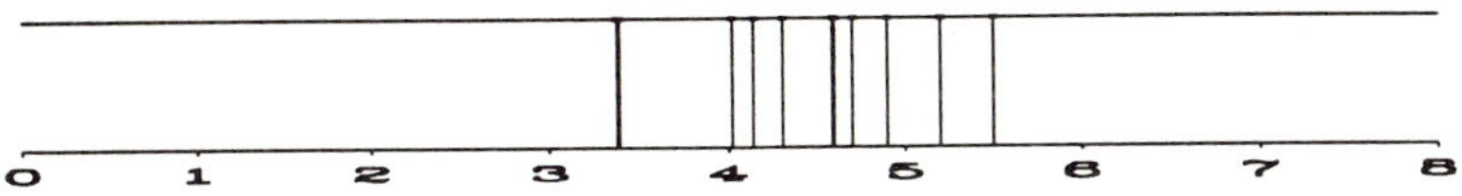

eq. 6 (Chiou 1985)

compound class: 13 chlorobenzenes (for 5 compounds BCF data on both fish species were used

species: Poecilia reticulata, Salmo gairdneri

log P_{ow}: 3.38 - 5.50

F

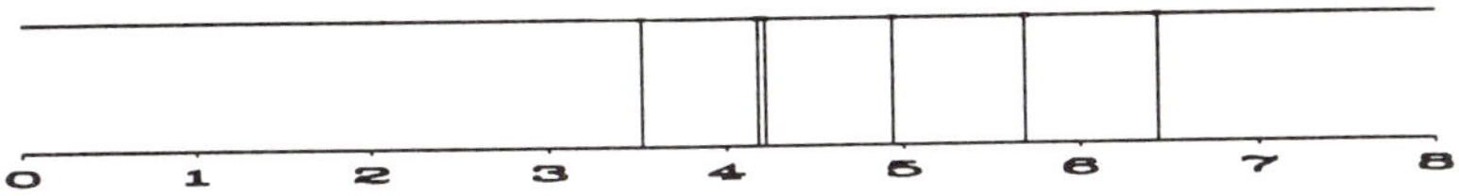

eq. 12 (Könemann and van Leeuwen 1980)

compound class: 6 chlorobenzenes

species: Poecilia reticulata

log P_{ow}: 3.53 - 6.44

G

eq. 13 (Connell and Hawker 1988)

compound class: 42 chlorinated hydrocarbons

species: Poecilia reticulata, Carassius auratus, Salmo gairdneri

log P_{ow}: 2.60 - 9.84

Figure 5.5 (continued) Assessment of the log P_{ow} parameter covered by BCF_{FISH}QSARs

solvent characteristics between natural lipids and 1-octanol. It appears plausible, that for molecules less than a certain volume or certain dimensions, 1-octanol is a satisfactory surrogate, i.e. the activity coefficients in 1-octanol and the fish lipid are approximately equal. But for larger molecules this similarity breaks down and the activity coefficients in the membrane phase are much larger than in 1-octanol, which then is no longer a satisfactory surrogate and log P_{ow} is no longer a satisfactory descriptor. It seems likely that the relatively low solubility of voluminous molecules in membranes as compared to 1-octanol is one of the causes for the loss in linear correlation between log P_{ow} and log BCF. In that case, membrane/water partition coefficients may be used as a more reliable parameter to estimate and correlate the BCFs of organic chemicals in aquatic organisms (Gobas et al. 1987).

5.6.3 Deviations due to Large Molecular Diameter

Relatively large molecules, such as many of the high P_{ow} compounds may experience a hindered membrane passage. From a mechanistic point of view it seems conceivable that a cut-off diameter exists for passive membrane diffusion. Lack of permeation was observed for lipophilic chemicals with effective cross sections over 9.5 Å, e.g. hepta- and octachloronaphthalenes (Opperhuizen et al. 1985). It was demonstrated from the relationship between the steric configuration of the compounds and the lack of uptake that also the type and composition of the organisms' membranes can influence the bioconcentration potential of a chemical. In agreement with the size limited uptake are results reported for disperse dyestuffs and pigments (Anliker et al. 1988) revealing no substantial bioconcentration.

5.6.4 Deviations due to Substructure Contributions

Dinitrophenols are reported to show less bioconcentration than estimated from log P_{ow} (Butte 1987). The deviation is not completely explainable by the compounds' considerable dissociation (descriptor: pKa). An indicator variable "2,4-dinitro-substitution" is needed, a constant factor has to be subtracted to describe BCFs in a heterogeneous series of phenols, which may account for the reduced redox-stability of these compounds (Hauk et al. 1990). In contrast to this finding, Deneer et al. (1987) reported m-dinitrobenzene to accumulate more than estimated from log P_{ow}. It has to be considered that log P_{ow} may not be the appropriate parameter to describe these compounds partitioning.

5.6.5 Deviations due to Degradation

Transformation processes mostly result in reduced lipophilicity, and hence reduced BCF values. For superlipophilic compounds even very low metabolic transformation rates can have a significant effect on the elimination rate, and hence BCF, as the elimination by diffusion is very low. Examples of degradation interfering with bioconcentration are the reduction of nitro-groups, or the rapid transformation of benzo(a)pyrene in fish (Spacie et al. 1983). Enhanced metabolism and low bioconcentration of polychlorinated dibenzo-p-dioxins in various fish species (Gobas and Schrap 1990), if assumed to be lipophilicity related, substantiate the diminished slope of a QSAR function explicitly for this class of compounds.

Disregarding the effect of biodegradation reducing bioconcentration for QSAR predictions with the scope of environmental hazard assessment results in estimates on the "safe side".

5.6.6 Outliers Revealing Apparent Excess Bioconcentration

Only few examples of QSAR estimated BCFs being considerably lower than experimental values have been reported, e.g. m-dinitrobenzene (Deneer et al. 1987) and diuron (Manthey et al. 1990). Mostly, outliers revealing excess bioconcentration were identified with respect to QSARs having relatively low slopes and high intercepts. It can now be speculated that the derived functions are misleading by being fitted to data points representing compounds of diminished bioconcentration potential. In that case, the "outliers" revealing apparent excess bioconcentration are points of the "worst case" function, whereas the majority of the analysed chemicals are "negative outliers". If this hypothesis holds true, the problem reduces to statistical considerations with respect to appropriate selection of the data set for deriving QSARs.

5.7 Conclusions

The major conclusion drawn from this investigation is: The most severe problems for predicting bioconcentration potential by QSARs are not posed by the inherent QSAR problems, such as the uncertainties associated with descriptors and statistics. In fact, the problems are predominantly due to the inconsistent experimental material any modeling effort is based on.

The estimation of BCF values from log P_{ow} is founded on a relatively profound theoretical basis. However, the predictive power of the QSARs should not be overestimated and their limitations realized.

The linear log P_{ow}/log BCF correlations only hold for compounds having log $P_{ow} < 6$. For predicting superlipophilic chemicals' BCFs, non-linear relationships have to be applied and in most of these cases rather a chemical by chemical evaluation should be carried out.

In principle, QSARs predict BCF values corresponding to the average accumulation potential observed with the class of compounds investigated, hence the estimates do not necessarily reflect the potential "worst case".

Still, it is possible to construct such a function, which is not a QSAR in the strict sense.

The respective estimates correspond to the highest accumulation potential associated with the assumed lipophilicity. The presented bilinear function formalizes the empirical rules for estimating log BCF from log P_{ow}: The bioconcentration potential is linearly correlated with lipophilicity in the log P_{ow} range 0 - 6 with then log BCF corresponding to log P_{ow}. Compounds of higher lipophilicity reveal no further increase in BCF. Additional factors resulting in diminished bioconcentration are not considered, the various contributions are not systematically accountable. This procedure can be justified as a realistic "worst case" approach being necessary for assessing environmental hazard.

Acknowledgements

The author thanks H. Wienen for support in data acquisition and Prof. Dr. W. Klein, Dr. S. Bradbury and J. DeBruijn for their helpful suggestions and stimulating discussions. This study was supported by the Federal Environment Ministry/Federal Environment Agency.

5.8 References

Anliker, R., Moser, P. and Poppinger, D. (1988). Bioaccumulation of dyestuffs and organic pigments in fish. Relationships to hydrophobicity and steric factors. *Chemosphere*, 17, 1631 - 1644.

Baughman, G.L. and Paris, D.F. (1982). Microbial bioconcentration of organic pollutants from aquatic systems-a critical review. *CRC Crit. Rev. Microbiol.* 205 - 227.

Bruggeman, W.A., Opperhuizen, A., Wijbenga, A. and Hutzinger, O. (1984). Bioaccumulation of superlipophilic chemicals in fish. *Toxicol. Environ. Chem.* 7, 173 - 189.

Butte, W., Willig, A., Zauke, G.P. (1987). Bioaccumulation of phenols in zebrafish determined by a dynamic flow through test. In: Kaiser, K. L. E. (Ed.) QSAR in environmental toxicology-II, D. Reidel, Dordrecht.

Bysshe, S.E. (1982). Bioconcentration factor in aquatic organisms. In: Handbook of chemical property estimation methods, Lyman. W. J. (Ed.) Mc Graw-Hill, New York.

Chiou, C.T. (1985). Partition coefficients of organic compounds in lipid-water systems and correlations with fish bioconcentration factors. *Environ. Sci. Technol.* 19, 57 - 62.

Collander, R. (1951). The partition of organic compounds between higher alcohols and water. *Acta Chem. Scand.* 5, 774 - 780.

Connell, D.W. (1988). Bioaccumulation behaviour of persistent organic chemicals with aquatic organisms. *Rev. Environ. Contam. Toxicol.* 101, 117 - 154.

Connell, D.W. and Hawker, D.W. (1988). Use of polynominal expressions to describe the bioconcentration of hydrophobic chemicals by fish. *Ecotoxicol. Environ. Saf.* 16, 242 - 257.

DeBruijn, J., Busser, F., Seinen, W. and Hermens, J.L.M. (1989). Determination of octanol/water partition coefficients for hydrophobic organic chemicals with the "slow-stirring" method. *Environ. Toxicol. Chem.* 8, 499 - 512.

Deneer, J.W., Sinnige, T.L., Seinen, W. and Hermens, J.L.M. (1987). Quantitative structure-activity relationships for the toxicity and bioconcentration factor of nitrobenzene derivatives towards the guppy (Poecilia reticulata). *Aquat. Toxicol.* 10, 115 - 129.

Geyer, H., Sheehan, D., Kotzias, D., Freitag, D. and Korte, F. (1982). Prediction of ecotoxicological behaviour of chemicals: relationship between physicochemical properties and bioaccumulation of organic chemicals in the mussel. *Chemosphere* 11, 1121 - 1134.

Gobas, F.A.P.C., Shiu, W.Y. and Mackay, D. (1987). Factors determining partitioning of hydrophobic organic chemicals in aquatic organisms. In: Kaiser, K.L.E. (Ed.) QSAR in environmental toxicology-II. D. Reidel, Dordrecht.

Gobas, F.A.P.C. and Schrap, S.M. (1990). Bioaccumulation of some poly chlorinated dibenzo-p-dioxins and octachlorodibenzofuran in the guppy (Poecilia reticulata). *Chemosphere* 20, 495 - 512.

Hauk, A., Richartz, H., Schramm, K.W. and Fiedler, H. (1990). Reduction of nitrated phenols: a method to predict half-wave potentials of nitrated phenols with molecular modeling. *Chemosphere* 20, 717-728.

Hawker, D.W. and Connell, D.W. (1986). Bioconcentration of lipophilic compounds by some aquatic organisms. *Ecotoxicol. Environ. Saf.* 11, 184 - 197.

Hawker, D.W. and Connell, D.W. (1989). A simple water/octanol partition system for bioconcentration investigations. *Environ. Sci. Technol.* 23, 961 - 965.

Hawker, D.W. (1990). Description of fish bioconcentration factors in terms of solvatochromic parameters. *Chemosphere* 20, 467 - 477.

Kamlet, M.J., Doherty, R.M., Carr, P.W., Mackay, D., Abraham, M.H. and Taft, R.W. (1988). Linear solvation energy relationships. 44. Parameter estimation rules that allow accurate prediction of octanol/water partition coefficients and other solubility and toxicity properties of polychlorinated biphenyls and polycyclic aromatic hydrocarbons. *Environ. Sci. Technol.* 22, 503 - 509.

Karickhoff, S.W., Brown, D.S. and Scott, T.A. (1979). Sorption of hydrophobic pollutants on natural sediments and soil. *Water Res.* 13, 241 - 248.

Kenaga, E.E. and Goring, C.A. (1980). Relationship between water solubility, soil sorption, octanol-water partitioning and bioconcentration of chemicals in biota. In: Eaton, J. G. et al. (Eds.) Aquatic Toxicology, Vol. 707, ASTM, Philadelphia.

Könemann, H. and van Leeuwen, R. (1980). Toxicokinetics in fish: accumulation and elimination of six chlorobenzenes in guppies. *Chemosphere* 9, 3 - 19.

Leahy, D.E., Taylor, P.J. and Wait, A.R. (1989). Model solvent systems for QSAR part I. Propylene glycol dipelargonate (PGDP). A new standard solvent for use in partition coefficient determination. *Quant. Struct.- Act. Relat.* 8, 17 - 31.

Mackay, D. (1982). Correlation of bioconcentration factors. *Environ. Sci. Technol.* 16, 274 - 276.

Manthey, M., Smolka, S., Faust, M., Bödecker, W. and Grimme, L.H. (1990). Relationships between lipophilicity, bioaccumulation and toxicity of phenylureas in Chlorella fusca. Poster presented at the 7th International Congress on Pesticide Chemistry, IUPAC, Hamburg, August 5 - 10, 1990.

MedChem (1989). clogP Software Release 3.54.

Neely, W.B., Branson, D.R. and Blau, G.E. (1974). Partition coefficients to measure bioconcentration potential of organic chemicals in fish. *Environ. Sci. Technol.* 8, 1113 - 1115.

Oliver, B.G. and Niimi, A. (1983). Bioconcentration of chlorobenzenes from water to rainbow trout: correlation with partition coefficients and environmental residues. *Environ. Sci. Technol.* 17, 287 - 291.

Oliver, B.G. (1984). The relationship between bioconcentration factor in rainbow trout and physical-chemical properties for some halogenated compounds. In: Kaiser, K. L. E. (Ed.) QSAR in environmental toxicology. D. Reidel, Dordrecht.

Opperhuizen, A., Velde, E.W., Gobas, F.A., Lem, D.A. and Steen, J.M. (1985). Relationship between bioconcentration in fish and steric factors of hydrophobic chemicals. *Chemosphere* 14, 1871 - 1896.

Opperhuizen, A., Serné, P. and Van der Steen, J.M.D. (1988). Thermodynamics of fish/water and octan-1-ol/water partitioning of some chlorinated benzenes. *Environ. Sci. Technol.* 22, 286 - 292.

Ogata, M., Fujisawa, K., Ogino, Y. and Mano, E. (1984). Partition coefficients as a measure of bioconcentration potential of crude oil compounds in fish and shellfish. *Bull. Environ. Contam. Toxicol.* 33, 561 - 567.

Sabljic, A. (1987). Nonempirical modeling of environmental distribution and toxicity of major organic pollutants. In: Kaiser, K. L. E. (Ed.) QSAR in environmental toxicology-II. D. Reidel, Dordrecht.

Schrap, S.M. and Opperhuizen, A. (1990). Relationship between bioavailability and hydrophobicity: reduction of the uptake of organic chemicals by fish due to the sorption on particles. *Environ. Toxicol. Chem.* 9, 715 - 724.

Schüürmann, G. and Klein, W. (1988). Advances in bioconcentration prediction. *Chemosphere* 17, 1551 - 1574.

Smith, A.D., Barath, A., Mallard, C., Orr, D., McCarthy, L.S. and Ozburn, G.W. (1990). Bioconcentration kinetics of some chlorinated benzenes and chlorinated phenols in american flagfish, Jordanella floridae (Goode and Bean). *Chemosphere* 20, 379-386

Spacie, A. and Hamelink, J.L. (1982). Alternative models for describing the bioconcentration of organics in fish. *Environ. Toxicol. Chem.* 1, 309 - 320.

Spacie, A., Landrum, P.F. and Leversee, G.J. (1983). Uptake, depuration, and biotransformation of anthracene and benzo(a)pyrene in bluegill sunfish. *Ecotoxicol. Environ. Saf.* 7, 330 - 341.

Umweltbundesamt (1990). Grundzüge der Bewertung von neuen Stoffen nach dem ChemG. Umweltbundesamt, Berlin.

Veith, G.D. and Kosian, P. (1983). Estimating bioconcentration potential from octanol/water partition coefficients. In: Mackay, D. et al. (Eds.) Physical behaviour of PCBs in the Great Lakes. Ann Arbor Science Publishers, Ann Arbor.

Bioconcentration and Biomagnification: is a Distinction Necessary?

Antoon Opperhuizen

6.1 Abstract

Organic micropollutants can be taken up by aquatic organisms from water, food or sediments. To be able to assess which of the three routes is important for specific organisms and chemicals, knowledge of the mechanisms and kinetics of the various uptake processes is required.

A simple first order bioaccumulation model is described which allows one to assess the relative importance of the three routes of uptake for different chemicals in various organisms. During aqueous and dietary exposure the efficiency of uptake of organics from the water by the gill and from food via the intestines are both approximately 50 %.

Exposure to low concentrations in water results in a high uptake via the gill because aquatic organisms ventilate large volumes of water to satisfy their oxygen requirements.

The concentrations of hydrophobic contaminants in sediment and food are usually orders of magnitude higher than those in water. However, because the water ventilation rates (mL/g/day) of aquatic organisms are also several orders of magnitude higher than the ingestion rates of food and sediments (g/g/day) it is likely that for most organic chemicals water is the major source for bioaccumulation. Only for extremely hydrophobic chemicals uptake from food becomes important. Uptake from sediments is probably unimportant for aquatic species which are unable to digest sediments. For organisms which are able to feed on sediments, however, uptake via this route may significantly contribute to the total bioaccumulation.

The contributions of bioconcentration and biomagnification to the total bioaccumulation are dependent on both the properties of the chemical and the nature of the organism. With increasing hydrophobicity of the chemicals, it is likely that the contribution of biomagnification increases. The relative importance of biomagnification also increases with increasing size of the organisms. Bioaccumulation of extremely hydrophobic chemicals in predator organisms may thus be underestimated if only bioconcentration factors are used.

In order to make worst-case estimates of bioconcentration factors of different classes of organic chemicals in aquatic organisms, the use of aqueous activity coefficients rather than octan-1-ol/water partition coefficients is advocated.

[authors note: parts of this contribution have previously been published in Opperhuizen (in press); Opperhuizen and Schrap 1988 a; Opperhuizen and Sijm 1990]

6.2 Introduction

As part of the industrial and economic development, the production and use of synthetic chemicals has rapidly grown during the last few decades. Many thousands of chemicals are industrially produced nowadays, and many hundreds of new chemicals are registered each year. Most of these commercially produced chemicals are released in the environment and many of them can be taken up by organisms.

Many persistent organic chemicals which are of environmental concern such as chlorinated benzenes and biphenyls are routinely found in samples from the natural aquatic environment (Andersson et al. 1988). In particular in samples of sediments and aquatic biota significant amounts of such micropollutants have been detected. The concentrations of these chemicals in sediment and in biota are usually much higher than those found in the water, which is obviously due to their hydrophobicity. The low aqueous solubility, in combination with the organophilic properties of the compounds make that bioaccumulation and sediment sorption are important processes which determine the fate of these chemicals.

Relationships have been established between biota/water partition coefficients and hydrophobic parameters. For instance, good correlations have been found between bioconcentration factors (K_c) in fish (i.e. fish/water partition coefficients during steady state) and octan-1-ol/water partition coefficients ($K_{d,oct}$) of chlorinated benzenes, biphenyls and other structurally related chemicals (Bruggeman et al. 1984; Opperhuizen et al. 1985). In addition, sorption of organic micropollutants on sediment also seems to correlate with hydrophobicity of the chemicals (Karickhoff 1984). However, regression of sorption data with octan-1-ol/water partition coefficients has been less successful, which is partly due to the lack of reliable experimental sorption data.

Chemicals will accumulate in organisms if the rate constants for the processes of uptake are larger than those for the processes which control the elimination. During steady state or equilibrium the concentrations of this chemical in the organisms will then be higher than that in the surrounding environment.

A compound can be taken up by aquatic organisms from water, food or sediment. The process of bioaccumulation of organic chemicals by aquatic organisms may be summarized as:

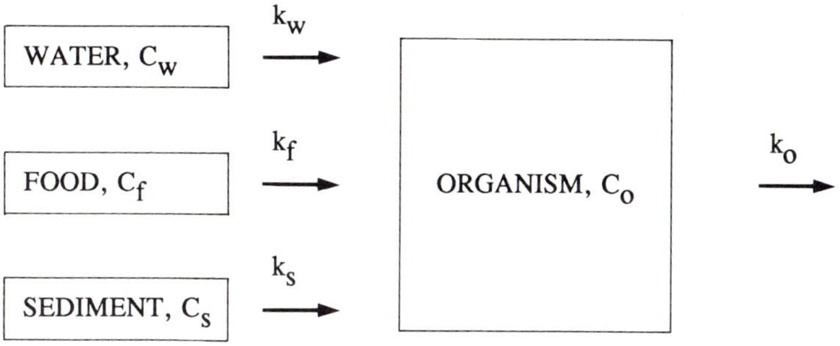

The change in the concentration of the chemical in the organism (C_o) over time (t) can be expressed by:

$$d\,C_o/dt = (k_w C_w + k_f C_f + k_s C_s) - k_o C_o \qquad (1)$$

Here C refers to concentration, k to a rate constant, and the subscripts w, f, s and o to water, food sediment and organism respectively. Equation 1 can also be written as:

$$d\,C_o/dt = (V_w E_w C_w + F_f E_f C_f + S_f E_s C_s) - k_o C_o \qquad (2)$$

In this equation V_w refers to the volume of water passing the gills, F_f to the amount of food and S_f the amount of sediment transported through the intestines per gram fish each day. The E values refer to the efficiencies of uptake of the chemical by the fish from the different media.

6.3 Bioconcentration: Uptake from Water

During the last two decades many studies have been published in which the results from bioconcentration experiments are reported. In most of these studies only bioconcentration factors have been listed. The bioconcentration factor of a chemical is the ratio of its concentrations in the organism and in water during steady state or equilibrium. One chemical usually has different biocencentration factors in different organisms. Even if the concentration of the chemical in organisms is expressed on lipid weight rather than on wet weight, bioconcentration factors usually differ significantly between various organisms. In particular for chemicals which can be biotransformed, significant species differences are observed.

In each organisms different bioconcentration factors (K_c) are normally found for different chemicals. In many studies relationships have been reported between bioconcentration factors and structure parameters or physico chemical properties of organic chemicals. In particular octan-1-ol/water partition coefficients (log K_d,oct) are often used to describe and predict bioconcentration factors. The use and application of octan-1-ol/water partition coefficients as a model for bioconcentration factors is widespread. Unfortunately little attention is usually paid to the fact that good correlations between log K_d,oct and log K_c have been found only for a few classes of chlorinated aromatic hydrocarbons, in particular for chlorinated benzenes, -biphenys, -naphthalenes and diphenylethers. For most other types of chemicals very poor correlations are often found. For chemicals such as silicones or nitrogen-containing aromatics and several other classes of organic chemicals, the octan-1-ol/water partition coefficients could not successfully be correlated with bioconcentration factors. So, in general it is not justified to expect that the octan-1-ol/water partition coefficient is a good model for organism/water partitioning of an organic chemical.

That for several classes of chlorinated benzenes and benzene-derivatives octan-1-ol/water partition coefficients and bioconcentration correlate well, is illustrated in Figure 6.1. Bioconcentration factors shown in this figure are all determined with fish which have wet weights less than 5 gram. All data originate from our own laboratory, and were determined under comparable experimental conditions (Opperhuizen and Sijm 1990). For most of the chlorinated aromatic hydrocarbons the bioconcentration factors tend to increase with increasing hydrophobicity of the compounds. It may be clear, however, that bioconcentration factors can only be predicted from octan-1-ol/water partition coefficients within approximately one order of magnitude. That more precise predictions are not reliable is not only due to experimental limitations, but also to differences in the chemical and physico chemical properties of the different classes of compounds. More evidence for these differences has previously been presented (Opperhuizen et al. 1988).

In the literature it is often suggested that uptake rate constants tend to increase with hydrophobicity for chemicals with log $K_{d,oct}$ values less than 3. For chemicals which are more hydrophobic it can be suggested from the data shown in Figure 6.2, that the uptake rate constants are not varying with a change in hydrophobicity. The data shown here do not support the suggestion that for extremely hydrophobic chemicals the uptake rate constants drop with increasing hydrophobicity. The uptake rate constants for hydrophobic organic chemicals in the small fish is approximately 1000 mL/g/d. Since the volume of water which is ventilated

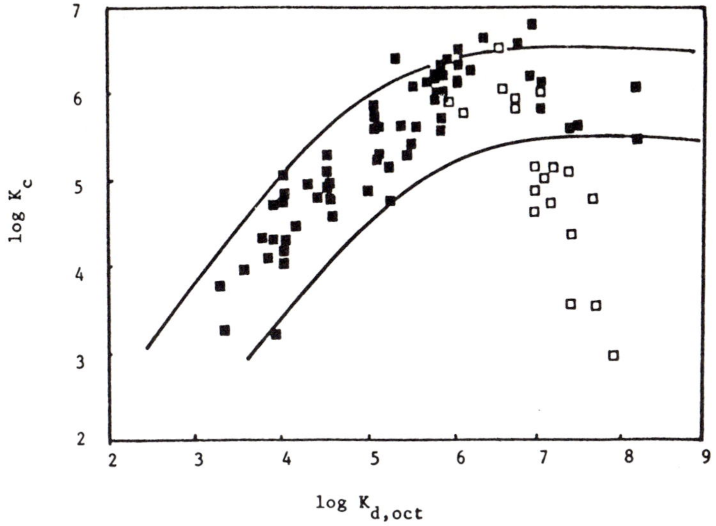

Figure 6.1 Relationship between log K_d,oct and log K_c of several classes of chlorinated aromatic hydrocarbons in small fish. Open circles represent dibenzo-p-dioxin and dibenzofuran congeners. Data were previously published in Opperhuizen and Sijm (1990).

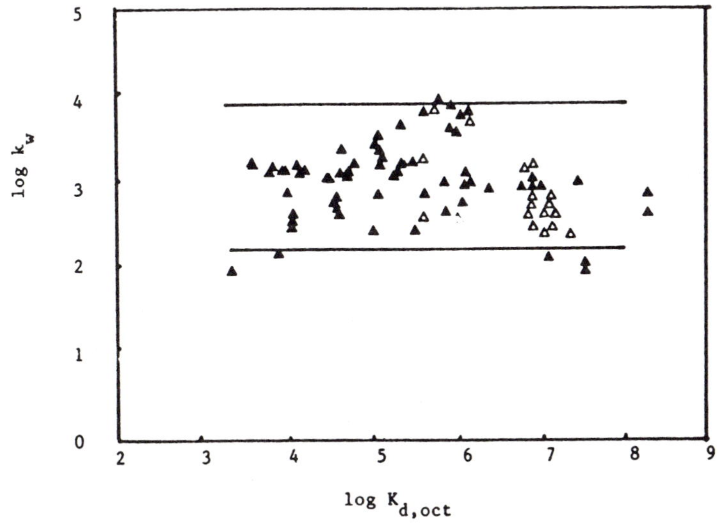

Figure 6.2 Relationship between log K_d,oct and log k_w of several classes of chlorinated aromatic hydrocarbons in small fish. Open circles represent dibenzo-p-dioxin and dibenzofuran congeners. Data were previously published in Opperhuizen and Sijm (1990).

across the gills is approximately 2000 mL/g/d, it may be argued that the gills extract approximately 50 % of the chemicals from the water (Opperhuizen and Schrap 1987). Comparable results have previously been published. For larger fish almost the same extraction efficiency has been reported. However, since the ventilation volume decreases significantly with increasing fish weight, it can be hypothesized that the uptake rate constants of chemicals will be lower for larger fish. This is supported by data reported in the literature. For fish with a wet weight of 100 gram the uptake rate constants are approximately 100 mL/g/d, while for large fish ($>$ 1 kg) the uptake rate constants are even lower.

In several bioconcentration experiments it has been observed that several extremely hydrophobic chemicals are not accumulated. In particular this has been observed for octachlorodibenzo-p-dioxin, octachlorodibenzofuran, octachloronaphthalene, hexabromobenzene and several other compounds. It has been proposed that these chemicals are not able to cross biological membranes by passive diffusion (Opperhuizen et al. 1985). All these chemicals have a cross section which is larger than 0.95 nm. Chemicals with comparable octan-1-ol/water partition coeffiecients or aqueous solubilities, such as octa- or decachlorobiphenyls, which have cross section less than 0.95 nm accumulate significantly in fish under the same experimental conditions.

In other experiments a lack of accumulation has been shown for hydrophobic chemicals with a length which exceeds 5.3 nm (Opperhuizen et al. 1987). This has been found for linear polydimethylsiloxanes in fish and with linear alkanes in rats. Recent data from a field study by Lamond et al. (personal communication) seem to support the hypothesis that chemicals with long chain lengths are poorly, if at all, taken up from water.

A model for the steric factors which may be important for passive diffusion of organic chemicals through membranes is presented in the Figures 6.3 and 6.4.

Many polychlorinated dibenzo-p-dioxins and dibenzofurans have bioconcentration factors which are significantly lower than those of polychlorinated biphenyls, -naphthalenes or diphenyl-ethers (Figure 1). For most PCDD and PCDF congeners this is not due to a small uptake rate constant, as can be seen in Figure 6.2. As is illustrated in Figure 6.5, many PCDDs and PCDFs have elimination rate constants which are much larger than those of the other chlorinated aromatic hydrocarbons. In several studies we have presented evidence that the relatively fast elimination should be explained by biotransformation (Opperhuizen and Sijm 1990). For chlorinated anisoles it has also been shown that biotransformation is of paramount importance for the overall rate of elimination. The rate constants of biotransformation of chlorinated aromatic hydrocarbons do not correlate with octan-1-ol/water partition coefficients. So, no relationships between log k_o or log K_c and log K_d,oct are found for chemicals which are significantly biotransformed. For several classes of chlorinated aromatic

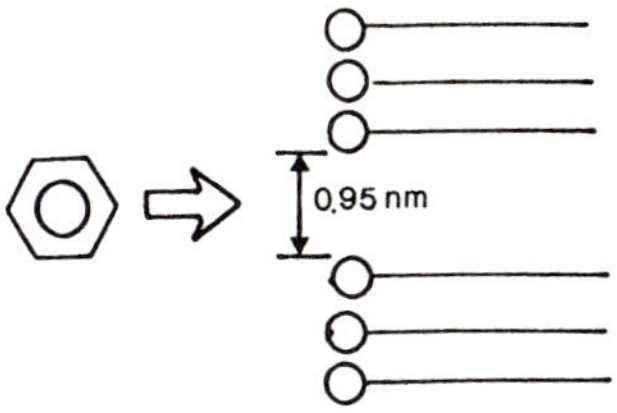

Figure 6.3 Schematic representation of steric properties of the chemicals and the membrane which may influence uptake of chemicals by organisms.

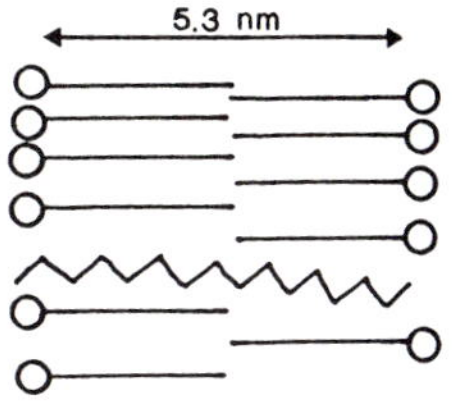

Figure 6.4 Schematic representation of steric properties of the chemicals and the membrane which may influence uptake of chemicals by organisms.

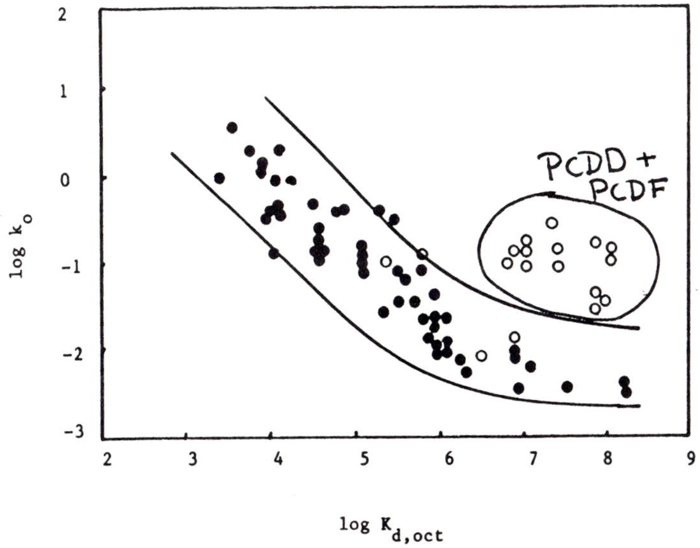

Figure 6.5 Relationship between log K_d,oct and log k_o of several classes of chlorinated aromatic hydrocarbons in small fish. Open circles represent dibenzo-p-dioxin and dibenzofuran congeners. Data were previously published in Opperhuizen and Sijm (1990).

hydrocarbons for which biotransformation is less important, the elimination rate constants tend to decrease with increasing hydrophobicity (Figure 6.5). Predicting elimination rate constants of chlorinated aromatic hydrocarbons from octan-1-ol/water partition coefficients is thus limited to chemicals for which biotransformation is negligible. However, even if biotransformation is unimportant, predictions will not be better than within one order of magnitude.

6.4 Biomagnification: Uptake from Food and Sediment

In many publications it is suggested that high concentrations of organic micropollutants which are often found in organisms of higher trophic levels of aquatic food chains and food webs are mainly the result of uptake via food. With the help of equation 2, this hypothesis can be investigated for hydrophobic chemicals.

For small fish uptake efficiencies of several chlorinated aromatic hydrocarbons are plotted against log $K_{d,oct}$ in Figure 6. Again, all data originate from our own lab. Based on these data and on data which were previously reported for several other fish species (Opperhuizen and Schrap 1988) it can be hypothesized that the uptake efficiency does not change with increasing hydrophobicity. Only steric factors may cause that for several compounds significantly lower uptake efficiencies are observed. For most chemicals which are significantly taken up from food the uptake efficiency is approximately 50 %. This corresponds nicely with the mean digestibility of food components. It may thus be suggested that the uptake efficiency of a chemical from the food is mainly determined by the digestibility of the food.

It has been documented in the literature that for a variety of organisms maintenance feeding rates range from 1 to 5 % of their body weight per day. So, whereas ventilation volumes of fish may vary several orders of magnitude, differences in feeding rates are limited to a factor 5. In a series of experiments with small fish which were fed 2 % of their body weight each day, biomagnification factors have been determined for several chlorinated aromatic hydrocarbons (Figure 6.7). Due to the decrease of elimination rate constants with increasing hydrophobicity and the constancy of E_f, the biomagnification factors tend to increase with increasing log $K_{d,oct}$ of these compounds. An identical trend is shown for bioconcentration factors in Figure 6.1. It must be noted, however, that for the small fish the bioconcentration factor of a chemical is five orders of magnitude higher than the biomagnification factor. This is due to the fact that both for the uptake from water and from food the uptake efficiency is approximately 50 %. The ventilation volume, however, is 2000 mL/g/d, while the feeding rate is only 0.02 g/g/d. Thus according to equation 2, for these small fish, uptake from food contributes significantly if the concentration of the micropollutant in food is five orders of magnitude higher than the concentration in water. If it is assumed that the food of higher predator fish is also fish, than the bioconcentration factor of the pollutant for the prey fish should exceed 10^5. As can be seen from Figure 6.1, only extremely hydrophobic chemicals have such high bioconcentration factors. For all chemicals which have low bioconcentration factors the overall bioaccumulation will be dominated by uptake from the water. This holds true both for the prey and the predator fish. For larger predator fish uptake from food contributes to the total bioaccumulation for chemicals with lower log K_d,oct values than for smaller predator fish. This because larger fish have significantly lower ventilation volumes than smaller fish, while the feeding rates are almost equal. Thus if the ratio between V_w and F_f for an organism drops one order of magnitude, then uptake from food is also important for chemicals which have a ten times lower bioconcentration factor. So, the fact that biomagnification is more important for large aquatic predator organisms is largly due to the decrease of ventilation volumes.

In literature, data on the uptake of organic micropollutants from sediments are very scarce. To the best of our knowledge, no data on the uptake of chlorinated aromatic hydrocarbons by aquatic organisms have been reported which can be compared with the data shown in the Figures 6.1, 6.2, 6.5, 6.6 and 6.7. Based on equation 1, it may be speculated for which chemicals and organisms uptake from sediments may be important. To allow an assessment two extreme cases may be considered.

<u>First</u>, organisms may be able to digest sediments, or parts of it. In this case, sediment acts as food and the total amount of chemical which may be taken up from the sediment is dependent on the rate of sediment ingestion and on the uptake efficiency. If the sediment (or parts of it) is an ideal food source, than the uptake efficiency will be determined by the digestibility of the sediment. The digestibility does, however, not only determine the uptake efficiency, but it will also determine the sediment ingestion rate. If the sediment is well digested, than the uptake efficiency will be high, but the ingestion rate will be low. On the other hand, if the sediment is poorly digested, the ingestion rate should be high to satisfy the nutrient and energy requirements of organisms which feed on the sediment, but the uptake efficiency will be low.

<u>Second</u>, organisms may be unable to digest any part of the sediment. In this case the uptake of the chemical by the organism from the sediment is determined by the desorption rate constant of the chemical from the sediment and the sediment residence time in the intestines. If it is assumed that the sediment ingestion rate is very high, than the desorption rate is determining the uptake efficiency. Recently desorption rate constants of several chlorinated benzenes and biphenyls have been reported (Brusseau et al. 1990). It was observed that desorption rate constants tend to decrease with increasing sorption coefficients. For sediment, it was found that if the sorption coefficients exceed a value of 10^4, the desorption rate constants are lower than $0.1\ d^{-1}$. So, if an organism which has a ventilation volume of 2000 ml/g/d is ingesting

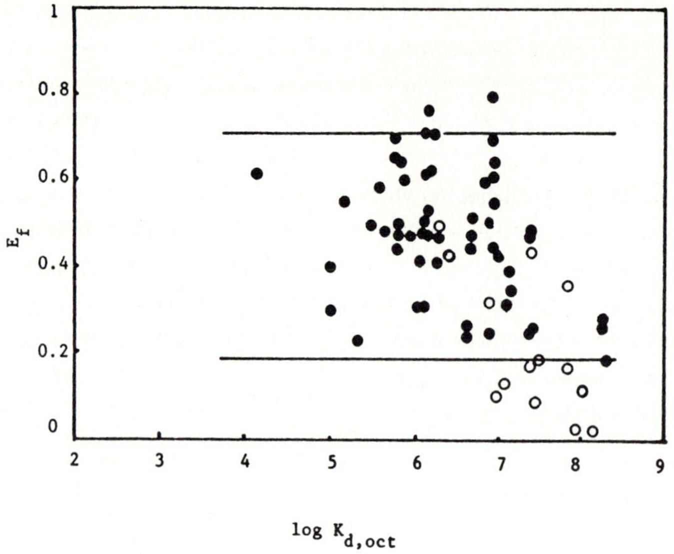

Figure 6.6 Relationship between log K_d,oct and E of several classes of chlorinated aromatic hydrocarbons in small fish. Open circles represent dibenzo-p-dioxin and dibenzofuran congeners. Data were previously published in Opperhuizen and Sijm (1990).

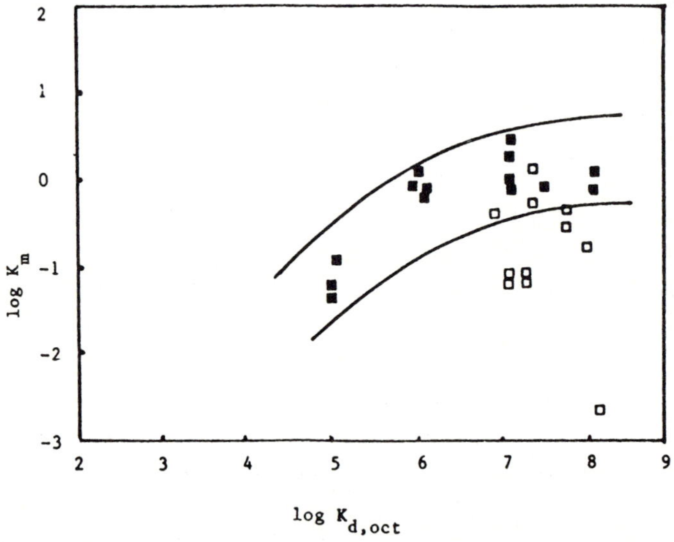

Figure 6.7 Relationship between log K_d,oct and log K_m of several classes of chlorinated aromatic hydrocarbons in small fish. Open circles represent dibenzo-p-dioxin and dibenzofuran congeners. Data were previously published in Opperhuizen and Sijm (1990).

sediment with a rate as high as 0.2 g/g/d, the ratio between V_w and S_f is 10^4. Uptake from the sediment which has a concentration of the chemical which is 10.000 times higher than that in water, is then as important as uptake from water if the uptake efficiencies E_w and E_s are equal (i.e. $V_w E_w C_w = S_f Es C_s$). This, however, will not be the case. Whereas E_w will be 0.5, E_s will be lower than 0.1 if the rate constant for desorption is 0.1 d^{-1} and the sediment can not be digested. For chemicals which are more hydrophobic, the sorption coefficients may be higher than 10^4, but the desorption rate constants will consequently be smaller. Although the rates of desorption may be slightly higher in the intestines of organisms, and the water ventilation volumes may be lower for several aquatic organisms, it seems unlikely that direct uptake of micropollutants from sediments contributes significantly to overall accumulation for most organisms. Only for organisms which are able to digest parts of the sediment this route of uptake may play a role in the accumulation of extremely hydrophobic chemicals which have high sediment sorption coefficients.

6.5 Worst Case Assessment of Bioaccumulation

The relationship between bioconcentration factors and octan-1-ol/water partition coefficients (Figure 6.1) seems to have an upper limit. To allow 'worst case' assessment of biotic effects it is important to estimate the 'maximum' bioconcentration factors. A simple estimation method which may be helpful for regulatory purposes is to assess the bioconcentration factor as being the activity coefficient of the chemical in water.

The rationale behind this method is that a partition coefficient of a chemical between an organic solvent and water can be expressed as

$$K = \gamma w / \gamma o$$

Here γw and γo are the activity coefficients of the chemical in water and organic solvent respectively. Ideal solubility of an organic pollutant in an organic solvent can be represented by γo being unity. In that case the partition coefficients K equals the activity coefficient of the chemical in the water phase.

The aqueous activity coefficient is calculated from the aqueous solubility corrected for the melting point. So, if a hydrophobic chemical 'dissolves' ideally in the lipids of an aquatic organisms (γlipid = 1), then the lipid normalized bioconcentration factor can not exceed the aqueous activity coefficient. In various studies it has been shown, however, that organic micropollutants do not resemble lipids. The pollutants activity coefficients in the fish lipid phase are thus not unity. Worst case bioconcentration factors may, however, be estimated with the ideality assumption. In Figure 6.8 this is shown for the chemicals for which their

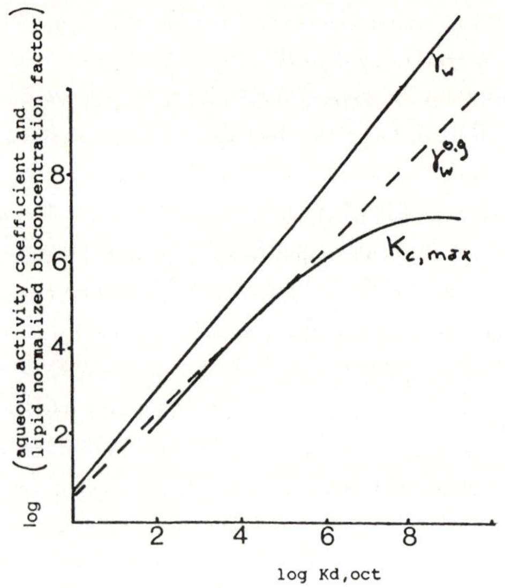

Figure 6.8 Relationships between octan-1-ol/water partition coefficients of hydrophobic chemicals and aqueous activity coefficients and lipid-normalized bioconcentration factors.

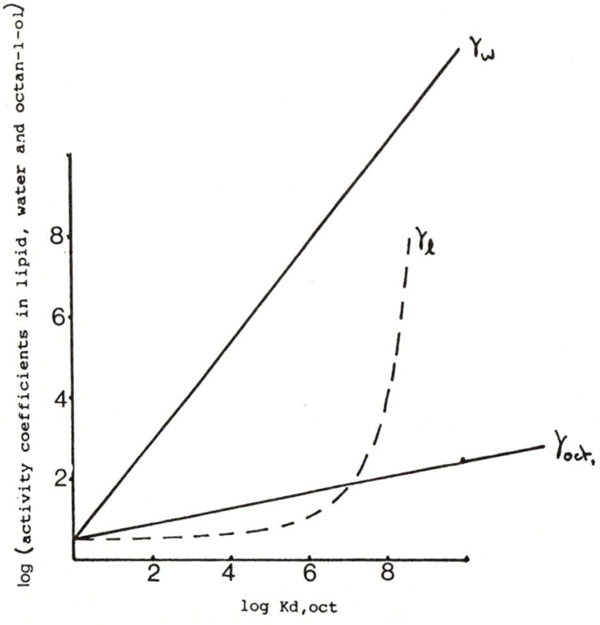

Figure 6.9 Relationship between octan-1-ol/water partition coefficients of hydrophobic chemicals and apparent activity coefficients in water, octan-1-ol and fish-lipids.

bioconcentration factors were plotted in Figure 6.1. In Figure 6.8, γ_w and log Kc are plotted against their octan-1-ol/water partition coefficients. The upper limit of the lipid normalized bioconcentration factors, is represented by the curve Kc,max. We have also shown the curve of $\gamma_w^{0.9}$. Since none of the bioconcentration factors is above this curve, the equation $Kc < \gamma_w^{0.9}$, may thus be used to make worst estimates of bioconcentration factors.

With the help of equation 3 and the original data of Figure 6.1, apparent activity coefficients of the test compounds for fish lipids and octan-1-ol are calculated. As is shown in Figure 6.9, the apparent activity coefficient in octan-1-ol increases with increasing hydrophobicity. Comparable results have previously been reported in the literature. It is remarkable that the lower limit of the fish lipids apparent activity coefficients of the test compounds is lower than that for octan-1-ol. The second remarkable observation is a sharp increase of the lower limit of fish lipid apparent activity coefficients for extremely hydrophobic chemicals. A variety factors may cause this dramatic increase, one of them being a steric factor related to the size of the test compounds relative to the dimensions of lipid membranes.

6.6 Conclusions

Neutral organic chemicals are rapidly taken up from water or food by aquatic organisms. For hydrophobic chemicals the efficiencies of uptake are not dependent on the hydrophobicity. However, organic chemicals which have a cross section which exceeds 0.95 nm or a length which exceeds 5.3 nm are taken up slowly, if at all.

The rate constant for elimination tend to decrease with increasing hydrophobicity for organic chemicals which are not significantly biotransformed. Biotransformation rate constants do not correlate with physico chemical properties.

Predictions of bioconcentration and biomagnification factors with octan-1-ol/water partition coefficients are usually not better than within one order of magnitude. For chemicals which are not chlorinated aromatic hydrocarbons predictions may be even worse. Worst case estimates of bioconcentration factors can be made from the reciprocal of supercooled liquid aqueous solubilities.

For most organic micropollutants which bioaccumulate uptake from water is the predominant route of uptake. Only for extremely hydrophobic organic chemicals uptake from contaminated food contributes significantly to the total bioaccumulation. Uptake from sediments will probably not contribute significantly to the total bioaccumulation in aquatic species. Only chemicals which have very high sorption coefficients may be taken up by organisms which are able to digest (parts of) the sediment. In spite of the fact that direct uptake from sediment may not be important for most aquatic organisms a constant ratio between the concentrations of pollutants in biota and in sediment may be found, due to steady state conditions in water, sediment, and organisms.

6.7 References

Andersson, O., Lindner, C.E., Olsson, M., Reuthergardh L., Uvemo, U.B. and Windquist, U. (1988), *Arch. Environ. Contam. Toxicol.*, 17, 755-765

Bruggeman, W.A., Opperhuizen, A., Wijbenga, A. and Hutzinger, O., (1984), *Toxicol. Environ. Chem.*, 7, 173-189.

Brusseau, M.L., Jessup, R.E. and Rao, P.S.C. (1990), *Environ. Sci. Technol.* 24, pp 727 -735.

Karickhoff, S.W. (1984), *J. Hydraul. Eng.* 19, 90-96.

Opperhuizen, A. (in press), Bioaccumulation kinetics: experimental data and modelling, In proceedings of the COST workshop, held in Lissabon, Portugal 1990, Fate of organic micropollutants in the aquatic environment.

Opperhuizen, A., Damen, H.W.J., Asyee, G.M., Van der Steen, J.M.D. and Hutzinger,O., (1987), *Toxicol. Environ. Chem.*, 13, 265 - 285.

Opperhuizen, A. and Schrap, S.M. (1987), *Environ. Toxicol. Chem.* 6, pp 335 - 342.

Opperhuizen, A. and Schrap, S.M. (1988), *Chemosphere* 17, pp 253 - 262

Opperhuizen, A. and Schrap, S.M. (1988), Relationships between sediment/water, fish/water and octan-1-ol/water partition coeffcients, In Conference Proceedings of the 1st European conference on ecotoxicology, October 1988, Copenhagen Denmark, H. Lokke, H. Tyle and F. Bro-Rasmussen eds. pp 480-489.

Opperhuizen, A., Serne, P., Van der Steen, J.M.D. (1988), *Environ. Sci. Technol.* 22, pp 286 - 292.

Opperhuizen, A. and Sijm, D.T.H.M., (1990), *Environ. Toxicol. Chem.* ,9, pp 175 - 186.

Opperhuizen, A., Van der Velde, E.W., Gobas, F.A.P.C., Liem, A.K.D., Van der Steen, J.M.D. and Hutzinger, O. (1985), *Chemosphere* 10, 1871 - 1896.

Bioconcentration of Xenobiotics from the Chemical Industry's Point of View

Norbert Caspers and Gerrit Schüürmann

7.1 Introduction

The members of the German chemical industry consider the assessment of bioconcentration potential as an important element in the development of new products. In this paper bioconcentration is discussed with respect to de facto consequences for the industry in terms of experimental test design and theoretical approaches using the tool of QSAR (Quantitative Structure-Activity Relationship).

The following questions will be dealt with in detail in the subsequent chapters:

7.2 What evidence do BCFs provide with regard to a realistic hazard assessment of xenobiotics?
7.3 What are the consequences when high BCFs are ascertained in chemical products?
7.4 Legal requirements
7.5 What practical test guidelines are available?
7.6 What time, costs or other expenditures do BCF (Bioconcentration Factor) measurements involve?
7.7 General comments on BCF prediction models
7.8 What is the current status of BCF prediction by QSARs with respect to several classes of compounds?

7.2 Informative Value of BCFs with Respect to a Hazard Assessment of Xenobiotics

As to this concern, the point of view of the chemical industry is not fundamentally different from that of the authorities and of the scientific community. Indeed, an experimentally assessed BCF normally does not provide evidence of an immediate risk, but definitely contributes to a risk potential. On the other hand, trying to evaluate BCF independently from other risk factors does not take into consideration the complex net of interacting effects which are exerted under real field conditions. A comprehensive hazard assessment of xenobiotics has to consider other risk factors as well such as toxicity, persistence and exposure including exposure kinetics and bioavailability. Particularly this latter factor is highly decisive for the ecological availability and hence for the environmental impact resp. the risk probability of chemical substances, in other words "the probability of occurrence of adverse effects" (terminology of the EC commission). Trying to define criteria for substances which are "dangerous to the environment", the EC Commission regards a combination of the factors toxicity, degradation and bioconcentration to be an appropriate means for classification and evaluation of industrial chemicals: e.g. BCFs higher than 100 are considered the more critical the lower their biodegradability is (even if acute toxicity to aquatic organisms is quite low).

Though in agreement with the basic idea of this EC proposal for industrial chemicals, the chemical industry is not the only body which looks critically upon some details of the suggested classification scheme (f.e. the proposed threshold values). The "German Advisory Committee on Existing Chemicals of Environmental Relevance" (BUA), a group consisting of representatives from science, the chemical industry and governmental authorities, is also afraid of the ever-increasing labelling of chemicals as "dangerous to the environment" which distracts attention from really hazardous substances of environmental relevance.

7.3 Consequences of high BCFs - Example: PCBs

Beyond the normal obligation to handle chemicals carefully, knowledge of high BCFs - in combination with other risk factors - may provoke further reactions from the authorities and the producers. The commercial polychlorinated biphenyls (PCBs) are a good instance of this type of consistent and uniform reaction from both groups in the F.R.G. as soon as knowledge on some undesirable features became available. Because of their physical properties (high thermal stability, nonflammability, excellent electrical insulating properties) PCBs had been used for decades in industry as hydraulic fluids, flame retardants, dielectric fluids, and heat transfer fluids. Especially in branches of industry where a high level of safety at work and plant safety is required (petrochemical plants, mining industry) PCBs had been used in large quantities.

In the 1960s and 1970s the scientific community became aware that PCBs tended to accumulate especially in higher trophic levels of (aquatic) food webs, facilitated by their persistence and lipophilic nature. Also effects on human health after chronic occupational exposure to PCBs were reported: hepatic dysfunction, decrease in some pulmonary functions, dermal toxicity etc. (Safe and Hutzinger, 1987; Deutsche Forschungsgemeinschaft, 1988). Moreover there was an initial suspicion that pyrogen decomposition of PCBs might lead to the formation of highly toxic polychlorinated dibenzofuranes (PCDF) and polychlorinated dibenzodioxins (PCDD). Finally the "German MAK Commission Liaison" classified the PCBs as "substances with suspected carcinogenic potential".

As a direct consequence of this increasing evidence the following actions were taken:

1972 Voluntary stop of sale for the use of PCBs in open systems by the Bayer AG, the only German producer

1976 Council Directive 76/769/EEC of July 1976 on the approximation of the laws, regulations and administrative provisions of the Member States relating to restrictions on the marketing and use of certain dangerous substances and preparations. In this Directive restraints on sale were ordered and the use of PCBs was permitted only in closed systems.

1978 Implementation of Directive 76/769/EEC as a national law of the F.R.G. (10. Verordnung zur Durchführung des Bundes-Immissionsschutzgesetzes)

1983 Voluntary withdrawal from the market by the Bayer AG

1985 Council Directive 85/467/EEC of 1 October 1985 amending the sixth time (PCBs/PCTs) Directive 76/769/EEC on the approximation of the laws, regulations and administrative provisions of the Member States relating to restrictions on the marketing and use of certain dangerous substances and preparations. In this Directive the use of PCBs even in closed systems was also forbidden.

1989 Implementation of Directive 85/467/EEC as a national law of the F.R.G. (PCB-, PCT-, VC-Verbotsverordnung)

The voluntary stop of PCB production and sale is in accordance with the policy guidelines for environmental protection and safety at Bayer which were published in 1986 ("... Where health or environmental considerations demand it, the sale of products is curtailed or their production halted, regardless of economic interests. The measures deemed to be necessary on the basis of scientific knowledge are implemented in agreement with the authorities and the "Employers Liability Insurance Association" in Germany...").

The example of PCBs touches another problem which arises when the sale of products is halted (for whatever reasons): the problem of substitutes. On one hand, the potentially suitable substitutes are expected to possess better properties with respect to their environmental impact. On the other hand, the substitutes must make it possible to achieve the same standard of safety at work. Looking at the long list of potential PCB substitutes (Umweltbundesamt, 1989) one can see how difficult it is to combine both desirable features in one product. Obviously the only possible approach is to carefully consider the advantages and disadvantages of potential substitutes in every single case.

The basic substitutability of chemicals and the potential disadvantages in case of non-use must be thoroughly investigated before a ban is ordered.

7.4 Legal Requirements

The example of the PCBs evidently underlines the need for a thorough investigation of chemical substances with respect to the assessment of BCFs.

Within the F.R.G. the testing of biological and ecotoxicological properties of new products is regulated by the following laws and general directives:

Council Directive of 18 September 1979 amending for the sixth time Directive 67/548/EEC on the approximation of the laws, regulations and administrative provisions relating to the classification, packaging and labelling of dangerous substances (79/831/EEC)

German Chemicals Act: Gesetz zum Schutz vor gefährlichen Stoffen (Chemikaliengesetz) Bundesgesetzblatt 1990, Teil I, p. 521 - 547, vom 22.03.1980.
A test on bioconcentration in fish has to be performed on chemicals which reach level 1 (sales > 100 tonnes p.a. or total sales > 500 tonnes)

Directive of the Biological Research Centre for Agriculture and Forestry, F.R.G.: Directive for the official testing of crop protection agents. Part I: Instructions for application for the registration of pesticides (June 1987)

7.5 Test Guidelines

At present (December 1990) the following test guidelines are available for the experimental assessment of BCF:

7.5.1 OECD

OECD Guideline 305 A: Bioaccumulation: Sequential Static Fish Test
OECD Guideline 305 B: Bioaccumulation: Semi-Static Fish Test
OECD Guideline 305 C: Bioaccumulation: Degree of Bioconcentration in Fish
OECD Guideline 305 D: Bioaccumulation: Static Fish Test
OECD Guideline 305 E: Bioaccumulation: Flow-Through Fish Test,
all of them dating back to 1981.

According to oral information from the OECD, a comprehensive revision of these five test guidelines is envisaged; a detailed introductory paper to this revision developed by Takatsuki (Japan) will be circulated to the OECD member countries in the near future.

In addition to this new activity, OECD guideline 305 E has already been revised in recent years as part of the regular updating programme of the OECD. Currently it is still a draft which is not likely to be finally passed in the near future as there still exist some disagreements with respect to the comparability of BCFs derived from kinetics and those based on steady-state concentrations.

Of the five above mentioned test method alternatives, guideline 305 E and guideline 305 C are preferred by the German chemical industry for the determination of BCFs under flow-through conditions.

Normally the results of BCF determinations are based on whole-fish analysis, but portions of fish or specific organs may also be analysed if specific risks have to be considered.

According to guideline 305 E the test is only applicable to substances which are not readily degradable, relatively stable in water and soluble at less than 1 mg/l. The maximum length of the exposure phase is 28 days, followed by a depuration phase normally not longer than 56 days. For a given chemical, a rough estimate of the exposure and depuration phase via k_2 can be made using the relationships shown in Figure 7.1. It is important to note, however, that these equations are only valid within (pseudo-) 1st order kinetics, which latter is not always applicable for bioconcentration in fish (see, e.g. Nagel, 1988).

Following guideline 305 E, two concentrations should generally be tested, the higher concentration lying between one-tenth and one-hundredth of the acute LC50 of the substance, the second - if possible - one order of magnitude lower. The usually extremely low substance concentrations in the test medium - e.g. 10 - 40 ppb for some polychlorinated biphenyls - necessitate a specific and highly efficient chemical analysis: GC, HPLC, where necessary combined with radioisotopic techniques. When using radiolabelled material, it must be taken

Duration of exposure phase:

The time needed to reach 80 % or 95 % of the steady state can be estimated as

$$t_{80} = \frac{1.6}{k_2} \quad , \quad t_{95} = \frac{3.0}{k_2}$$

with k_2 = depuration rate constant [day $^{-1}$].

k_2 can be estimated using

$$\log k_2 = -0.414 \log K_{OW} + 1.47$$

$$\log K_{OW} = -0.862 \log S_W + 0.710$$

with K_{OW} = n-octanol/water partition coefficient

S_W = water solubility [mol/l]

Duration of non-exposure phase:

The half-life time to eliminate 50 % of the chemical is

$$t_{1/2} = \frac{0.693}{k_2}$$

In general, twice the duration of the exposure phase is regarded as being sufficient for the depuration phase.

Figure 7.1 Estimation of exposure and non-exposure phase of a bioconcentration test assuming (pseudo-) 1st order kinetics into account that (within the test period) readily metabolizable substances may produce considerable overestimations of the actual BCF. According to guideline 305 E, the exact chemical composition of potential metabolites need not be cleared up.

Besides a general description of the test design, the test report must give the steady-state BCF for each concentration of the test substance, the uptake rate constant and the depuration rate constant with their respective confidence limits.

OECD guideline 305 C is not based on the kinetic approach of guideline 305 E, which needs reliable information on the water solubility and the n-octanol/water partition coefficient of the test substance.

In OECD 305 C, basically only an exposure test is prescribed. Further differences to test guideline 305 E include:

test species: Orange-red killifish (Oryzias latipes) for the determination of TLm (median threshold limit resp. median tolerance limit), Japanese carp (Cyprinus carpio) for the determination of BCF

exposure levels: 1/1,000 and 1/10,000 of the 48-hour TLm duration of exposure phase: 56 days.

From the point of view of the German chemical industry, both OECD guidelines are regarded as acceptable prescriptions for the assessment of BCF. The existing draft of OECD guideline 305 E should be passed soon. Some of the OECD member countries, however, still see the need for further investigations before the revised guideline is finally published.

7.5.2 EC

EC Directive 79/831 "Accumulation - Flow Through Fish Test" is substantially based on OECD guideline 305 E; therefore a separate presentation is not necessary. Fundamental preparatory work for both guidelines was obtained from an EC ring test (Kristensen and Nyholm 1987), in which ecotoxicological laboratories from West Germany also participated.

7.5.3 EPA

For the registration of pesticides in foreign countries, especially in the North American market, the "EPA Pesticide Assessment Guidelines; Hazard Evaluation: Wildlife and Aquatic Organisms" (1982) are used. Depending on the actual field of application of the chemical under investigation, one of the following BCF test alternatives is required by the authorities:

§ 72-6 "Aquatic organism accumulation tests". Field of application: aquatic environment. Test parameter: kinetic approach; assessment of uptake rate constant and depuration rate constant.

§ 165-4 "Laboratory studies of pesticide accumulation in fish". Field of application: agriculture. Test parameter: assessment of bioconcentration and metabolism in edible and non-edible body parts of the test organisms.

The EPA alternatives deviate from OECD 305 E and EC 79/831 in the following respects:

the fixed exposure and depuration periods (28 and 14 days, respectively)
the testing of one concentration step only.

All the above-mentioned test guidelines are used for the assessment of BCFs in the laboratories of the German chemical industry.

7.6 Test Expenditure

The expenditure on BCF tests in the laboratories of the chemical industry cannot simply be expressed by the equation: time required x hourly rate = testing cost

Whereas the costs for the performance of the test in the biological laboratory may range from DM 10,000 - DM 50,000, the costs for the chemical analysis (including development of methods and quality assurance, perhaps the synthesis of labelled molecules and the use of radioisotopic techniques, in special cases also the identification of metabolites etc.) are usually much higher. As the time needed for analysis heavily depends on the physical and chemical properties and on the possibilities for detecting the test substance, exact figures cannot be given. Mean costs between DM 50,000 and DM 100,000 can be exceeded by far when particular technical problems are involved.

The question of whether the high costs involved will have a negative impact on innovation in the chemical industry will not be discussed here.

According to the German Chemicals Act, a test for species bioaccumulation of a new chemical product has to be performed only when the chemical reaches level 1; that means when the quantity of substance placed on the market reaches a level of 100 tonnes per year or a total of 500 tonnes. For many companies it may be advantageous from an economic point of view to charge an institute for applied scientific research with these occasional tests, rather than to obtain all the necessary equipment and trained staff themselves.

Besides these financial and organisational aspects, another kind of expenditure must be mentioned when reflecting upon biological testing in the Federal Republic of Germany. A lot of strict conditions have to be fulfilled before a laboratory can be issued with a license for testing living vertebrates (in accordance with the German animal protection law). Tests which are directly prescribed by national law or statutory order are simply notifiable. But all voluntary tests, including those which serve immediately as applicability tests for drafts or guidelines prior to their finalization (e.g. EC ring test "Bioaccumulation in Fish" 1984/85), are subject to approval by the regional authorities. Because of the rather complicated decision-making processes it usually takes several months to get (resp. not to get!) this approval.

Last but not least, there is an enormous amount of red tape confronting the notifier of new chemical products when strictly following the OECD principles of Good Laboratory Practice.

7.7 General Comments on BCF Prediction Models

QSAR methods are a theoretical tool for estimating the BCF of chemicals. A critical evaluation of these methods is of particular importance for the industry for the following two reasons.

On the one hand, the application of QSARs is becoming increasingly popular with regulatory authorities, and requests for further experimental testing are often based on lipophilicity data with the assumption that the latter correlate with bioconcentration. On the other hand, an in-house application of QSARs is meaningful only if the scope and current limitations of this approach are clearly identified. It should be mentioned that several companies have already started their own QSAR research on ecotoxicology in order to develop selection criteria for high-priority chemicals, and an important inter-company activity is also taking place in a working group of the VCI (Verband der Chemischen Industrie). One of the major reasons for these activities is the large number of existing chemicals with many data gaps with respect to ecotoxicological testing.

The lipophilicity-related interpretation of the bioconcentration phenomenon has been discussed thoroughly in the literature both from the pure QSAR standpoint as well as on the basis of thermodynamic considerations. So the scientific basis for linear log BCF-log K_{ow} correlationships appears to be well established. The fact that for some classes of compounds the BCF can be estimated within one order of magnitude by using adequate QSAR models is of great importance not only for the scientific communtity, but also for regulatory authorities and for the chemical industry. The latter is true because QSAR models now can be used as screening instruments both for existing chemicals and for newly developed compounds prior to experimental testing. At the same time, the more intensive use of QSAR models has shown that only expert application will lead to reliable results, because there are many instances where it is necessary to judge whether or not a QSAR model will be of help with the decision at hand.

7.8 Assessment of BCF Predictability using Several Classes of Chemical Compounds

In a recent analysis of the BCF-K_{ow} relationship (Schüürmann and Klein, 1988) it was shown that the previously claimed high-quality correlation between log BCF and log K_{ow} for a large variety of chemicals did not hold true for validated experimental log K_{ow} values according to the starlist of the THOR database. Instead, only a restriction to the subset of chlorinated and polyaromatic hydrocarbons (CHCs and PAHs) led to a highly significant log BCF-log K_{ow} QSAR model (correlation coefficient $r = 0.95$, $n = 22$, log K_{ow} range 2.0 - 6.5). The corresponding plot including two more compounds according to the updated list of experimental log K_{ow} data (THOR database Pomona 89) is shown in Figure 7. 2. It is

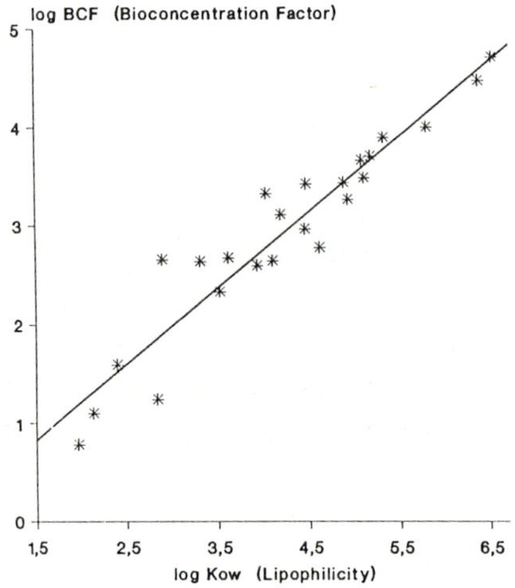

Figure 7.2 Relationship between log BCF and log K_{ow} for 24 chlorinated and polyaromatic hydrocarbons [data taken from Schüürmann and Klein (1988) augmented by two log K_{ow} values from Leo (1989)]. The stars denote the experimental log BCF values ordered by increasing experimental log K_{ow}, and the solid line represents the linear regression result.

seen that for this restricted set of 24 CHCs and PAHs the deviations in BCF from the regression line are within ca. half an order of magnitude, which already indicates that for more diverse compounds with varying reactivity properties a significant scatter is expected in a log BCF-log K_{ow} plot. Interestingly, for this subset of compounds pure geometric descriptors such as the solvent-accessible surface area and volume accounted equally well for the BCF variation (Schüürmann and Klein, 1988; Connell and Schüürmann, 1988).

On the other hand, it was found that for the total set of 49 organic compounds the correlation coefficient between log BCF and calculated log K_{ow} was only 0.63, which was partially due to the fact that the calculated log K_{ow} values according to Leo's scheme were not perfectly in line with the experimental counterparts (see Schüürmann and Klein, 1988 for details). This is of particular importance for the practical evaluation of the bioconcentration potential using computerised tools, because it is known that calculated log K_{ow} values according to Leo's scheme (Leo, 1989) though pretty reliable in general, may be in error for single molecules or even classes of compounds for more than two orders of magnitude (Schüürmann and Klein, 1988; Yen et al.,1989).

A good means of analysing the applicability of simple log BCF-log K_{ow} models is to compare for several classes of compounds the QSAR prediction line with the respective experimental values. For convenience, the above mentioned linear QSAR model for CHCs and PAHs according to (Schüürmann and Klein, 1988),

$$\log BCF = 0.78 \log K_{ow} - 0.35$$

was chosen as a reference line in order to identify systematic as well as random deviations of other compound classes with respect to CHCs and PAHs, which latter are in particular relatively stable compounds with no extraordinary molecular diameters. The regression equation is based on validated experimental log K_{ow} values according to the Starlist of the THOR database; interestingly, the regression parameters are very similar to those of Veith and Kosian's QSAR model based on 122 chemicals and 13 fish species (Veith and Kosian, 1983).

In Figure 7.3 a corresponding analysis is performed using literature data for polychlorinated biphenyls (Bruggeman et al., 1984; Gobas et al., 1989), top left in Figure 7.3), phenol derivatives part of which are chlorinated (Butte et al.,1987; top right), chloroanisoles (Opperhuizen and Voors, 1987; bottom left) and chloronitrobenzenes (Niimi et al., 1989; bottom right). Each plot contains the above mentioned linear QSAR model as a reference line, which is drawn in broken form in regions beyond the original log K_{ow} range of 2 to 6.5.

Concerning the polychlorinated biphenyls, it can be seen from Figure 7.3 that all but one (deca-chloro-biphenyl) of the seven congeners show considerably higher BCF values than

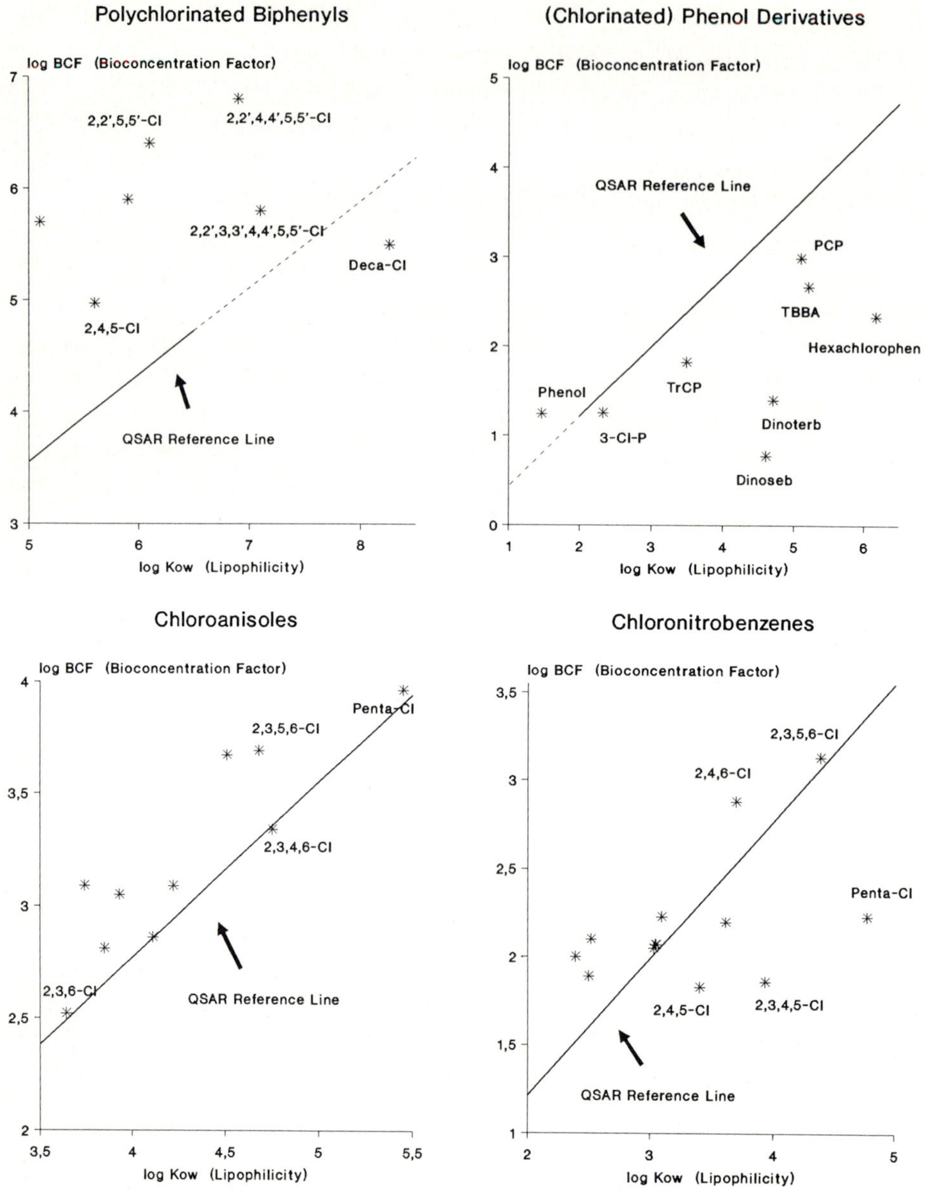

Figure 7.3 Comparison of the experimental log BCF-log K_{ow} data distributions with the linear QSAR model for CHCs and PAHs according to (Schüürmann and Klein, 1988) for four different classes of compounds. All experimental data are denoted by star symbols; the biphenyl data are taken from Bruggeman et al. (1984) and Gobas et al. (1989), the phenol data from Butte et al.(1987), the chloroanisole data from Opperhuizen and Voors (1987), and the chloronitrobenzene data from Niimi et al. (1989). Broken parts of the reference line indicate regions which exceed the log K_{ow} range (2 to 6.5) of the QSAR model. For convenience, some of the compounds in each of the plots are identified by abbreviated names.

would have been predicted by the simple linear model. This is in particular true for the first four compounds with log K_{ow} values less than 6.5, i.e. within the valid log K_{ow} range of the linear QSAR model. For example 2,2',5,5'-tetrachlorobiphenyl has a log BCF of 6.4 which is even larger than the respective log K_{ow} of 6.1. In particular, this BCF value is beyond the simplest upper limit estimate according to log BCF $\leq$ log K_{ow}, which would be assumed in the lipophilicity model of the bioconcentration process. The reason for the excess bioconcentration of this tetrachlorobiphenyl and other congeners is not clear apart from questions concerning the data quality of the experimental BCF and K_{ow} values. However, it is important to realize that for several chlorinated biphenyls the actual BCF data are ca. two orders of magnitude higher than would have been expected from the linear QSAR model of CHCs and PAHs. The decrease in bioconcentration in the series hexa-, octa- and decachlorobiphenyl can probably be attributed to an increase in molecular size (with increasing steric hindrance) and to a decrease in bioavailability because of the very low water solubilities of the higher congeners. On the whole, the data distribution shows that there are compounds with only little differences in log K_{ow}, but large differences in log BCF.

For the phenol derivatives, most of which are chlorinated, the situation is the reverse of the biphenyls: it can be seen from Figure 7.3 (top right) that here the linear QSAR model yields upper limits except for phenol. Similar to the chlorinated biphenyls, there are again compounds where the large differences in bioconcentration are not reflected in the lipophilicity scale. For example, dinoseb, dinoterb and pentachlorophenol are all in a log K_{ow} range of 4.6 to 5.1, but have BCF values which cover more than two orders of magnitude (log BCF from 0.8 to 3.0).

In contrast to both of these classes of compounds, the corresponding plot of the chloroanisoles in Figure 7.3 (bottom left) reveals a more systematic shift upwards with respect to the QSAR line, although even here there are isomers like the two tetrachloro-anisoles with almost identical log K_{ow} values but a difference in bioconcentration of one order of magnitude. However, on the whole the data pattern is in line with the general trend of increasing BCF values with increasing lipophilicity character of the compounds.

The last part of Figure 7.3 (bottom right) shows the corresponding data distribution for a series of chloronitrobenzenes. Inspection of the plot reveals that for this congeneric series of compounds the BCF variation is practically independent of lipophilicity: except for two outliers the data ensemble follows a line almost parallel to the log K_{ow} axis. Similar to the chloroanisoles there are again cases in which isomers of very similar log K_{ow} (here tri- and tetra-chloro-derivatives as indicated in Figure 7.3) differ in their BCF values by a factor of 10.

The present results based on literature data as shown in Figures 7.3 and 7.4 illustrate what

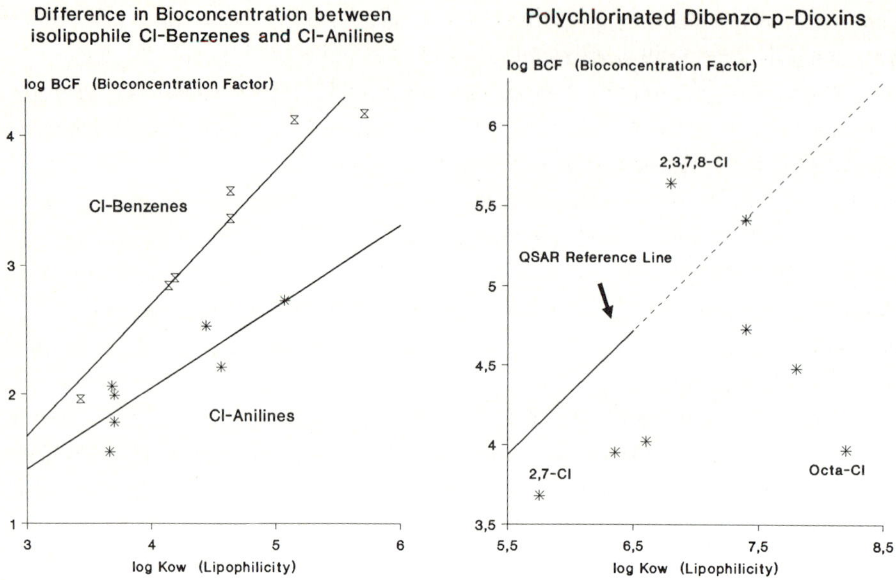

Figure 7.4 Comparison of the log BCF-log K_{ow} relationship for two series of compounds (left), and comparison of the relationship for dioxins with the QSAR model of Schüürmann and Klein (1988) (right). The data are taken from de Wolf et al. (1989) for the chlorobenzenes and chloroanilines, and from Gobas and Schrap (1990) for the dibenzo-p-dioxins, for details see legend of Figure 7.3

has been emphasized earlier (Schüürmann and Klein, 1988): The log BCF-log K_{ow} relationship is of less generality than is often assumed, but there are classes of compounds where this relationship in fact describes the structural dependence of the bioconcentration phenomenon adequately. The last two examples in this context are given in Figure 7.4, where in the left part chloro-benzenes and chloro-anilines are compared with respect to their log K_{ow}-dependence of log BCF (data taken from de Wolf et al., 1989). It is seen that for these two congeneric series there is a systematic difference in BCF for isolipophilic compounds (i.e. compounds with identical log K_{ow}), which increases with increasing liphophilicity (note the different slopes of the individual regression lines). This example again shows separate log BCF-log K_{ow} relationships for two classes of compounds; the different BCF values for isolipophilic compounds are attributed to differences in the biotransformation rates.

The last example on the right of Figure 7.4 shows a clearly non-linear log BCF-log K_{ow} data distribution for a series of polychlorinated dibenzo-p-dioxins (experimental data according to de Wolf et al., 1989). The maximum BCF is observed for the tetra-chloro-isomer (2,3,7,8-Cl as shown in the figure), and from this compound on the increase in log K_{ow} due to increased chlorine substitution leads to a significant decrease in BCF, covering a range of almost two orders of magnitude. This systematic trend which is the opposite of the simple linear QSAR model is not unexpected from the classical parabolic Hansch approach (Hansch, 1969), but apart from that simple geometric arguments such as the increased effective cross diameter of these fairly rigid compounds, for example, would at least partly explain the observed trend.

The consequences of this analysis for the necessary evaluation of the bioconcentration potential of compounds in the industry are summarized in the following statements and recommendations:

- For certain classes of compounds the predictive power of QSAR models is such that they can serve as reliable in-house screening tools for quantitative estimates of the bioconcentration potential both for existing and new chemicals.

- In particular, the relative influence of chlorine and alkyl substituents on the BCF potential of related compounds can be predicted if no specific metabolism is initiated at the same time.

- Apart from screening purposes the QSAR models for BCF are adequate tools for range-finding purposes in the context of experimental test design.

- For more complex structures, the log K_{ow} should be treated only as a rough indicator for the BCF order of magnitude; however, this may be significantly wrong with deviations in both directions, and appears to be no longer valid for log K_{ow} beyond 6.5.

- The use of calculated log K_{ow} values based on some standard scheme can be recommended in general, but must be approached with caution for more complex structures, where there are still errors of much more than one order of magnitude.

- The incorporation of QSAR models into decision-making schemes requires experts to judge the individual applicability for each compound under analysis; this is true both for in-house applications and for regulatory schemes.

- The complex phenomenon of bioconcentration requires more specific QSAR models in related expert systems which take into account other relevant characteristics like steric hindrance and biotransformation activity and other physico-chemical properties such as water and lipid solubility.

The last point is of particular importance also for the general applicability of the bioconcentration factor as an indicator of the bioconcentration potential. Compounds with extremely low water solubilities may show great BCF values but nonetheless may not reach critical limits in the organims, because the low saturation limit in both phases would just prevent them from entering the organism by passive diffusion in appreciable amounts. This aspect has not been adequately dealt with in the legislative decision-making schemes and in particular is not accounted for by current QSAR models.

7.9 Conclusions

In this paper the point of view of the German chemical industry on the bioconcentration of xenobiotics is presented; it is basically not different from that of the scientific community and the regulatory authorities. The existing draft of OECD guideline 305 E (not yet published), OECD guideline 305 C, and the test alternatives of the EPA are considered to be appropriate methods for the determination of BCFs. The potential maximum concentration (expressed as BCF) must be evaluated under consideration of exposure kinetics as well as uptake and depuration kinetics. The risk assessment has to link realistic bioaccumulated residues with possible effects. The expenditures (time, costs) for these determinations have grown rapidly in recent years primarily due to the highly sophisticated chemical analysis (GC, HPLC) in the test medium and in the organisms. The last two chapters of this paper deal with theoretical prediction models for the BCF. QSAR models for estimating BCF values of existing or newly developed chemicals must be applied with expertise, knowing the risk of erroneous results for different classes of compounds. On the other hand, the prediction of BCF values using only structure-related parameters offers a great opportunity for screening quickly for high bioconcentration potential; only systematic application of these techniques at a sophisticated level will show whether this approach will actually bring the expected benefits.

7.10 References

Bruggeman, W.A., Opperhuizen, A., Wijbenga, A. and O. Hutzinger (1984): Bioaccumulation of Super-Lipophilic Chemicals in Fish. *Toxicol. Environ. Chem.* 7, 173 - 189.

Butte, W., Willig, A. and G.-P. Zauke (1987): Bioaccumulation of Phenols in Zebrafish Determined by a Dynamic Flow Through Test. In: K. L. E. Kaiser (ed.): QSAR in Environmental Toxicology - II. D. Reidel Publishing Company, Dordrecht (NL), 43 - 53.

Connell, D.W. and G. Schüürmann (1988): Evaluation of Various Molecular Parameters as Predictors of Bioconcentration in Fish. *Ecotoxicol. Environ. Saf.* 15, 324 - 335.

Deutsche Forschungsgemeinschaft (1988): Polychlorierte Biphenyle. Bestandsaufnahme über Analytik, Vorkommen, Kinetik und Toxikologie - Mitteilung XIII der Senatskommission zur Prüfung von Rückständen in Lebensmitteln. VCH-Verlagsgesellschaft (Weinheim), 161 pp.

de Wolf, W., Yedema, E. and J. Hermens (1989): Bioconcentration of Aromatic Amines in Fish: A Possible Influence of Biotransformation. Poster presented at the 5th International symposium on responses of marine organisms to pollutants, April 12 - 14, Plymouth, UK.

Gobas, F.A.P.C. and S. M. Schrap (1990): Bioaccumulation of some polychlorinated Dibenzo-p-Dioxins and Octachlorodibenzofuran in the Guppy (Poecilia reticulata). *Chemosphere* 20, 495 - 512.

Gobas, F.A.P.C., Clark, K.E., Shiu, W.Y and D. Mackay (1989): Bioconcentration of Polybrominated Benzenes and Related Superhydrophobic Chemicals in Fish: Role of Bioavailability and Elimination into the Feces. *Environ. Toxicol. Chem.* 8, 231 - 245.

Hansch, C. (1969): A Quantitative Approach to Biochemical Structure-Activity Relationships. *Acc. Chem. Res.* 2, 232 - 239.

Kristensen, P. and N. Nyholm (1987): Bioaccumulation of chemical substances in fish, flow-through method. Final report of the ring test organized by the Water Quality Institute (VKI), Denmark under a contract with the Commission of the European Community.

Leo, A. (1989): CLOGP-3.54 MedChem Software. Daylight, Chemical Information Systems, Claremont, CA 91711, USA.

Nagel, R. (1988): Umweltchemikalien und Fische - Beiträge zu einer Bewertung. Habilitationsschrift, Universität Mainz. 256 pp.

Niimi, A.J., Lee, H.B. and G. P. Kossoon (1989): Octanol/Water Partition Coefficients and bioconcentration Factors of Chloronitrobenzenes in Rainbow Trout (Salmo Gairdneri). *Environ. Toxicol. Chem.* 8, 817 - 823.

Opperhuizen, A. and P. I. Voors (1987): Uptake and Elimination of Polychlorinated Aromatic Ethers by Fish: Chloroanisoles. *Chemosphere* 16, 953 - 962.

Safe, S. and O. Hutzinger (Edit.) (1987): Environmental Toxin Series 1. Polychlorinated Biphenyls (PCBs): Mammalian and Environmental Toxicology. Springer-Verlag (Berlin), 152 pp.

Schüürmann, G. and W. Klein (1988): Advances in Bioconcentration Prediction. *Chemosphere* 17, 1551 - 1574.

THOR database Pomona 89, MedChem Software 1989. Daylight, Chemical Information Systems, Claremont, CA 91711, USA.

Umweltbundesamt (1989): Ersatzstoffe für in Kondensatoren, Transformatoren und als Hydraulikflüssigkeiten im Untertagebergbau verwendete polychlorierte Biphenyle - Eine Zusammenstellung und Bewertung. Texte 33/89, 96 pp.

Veith, G.D. and P. Kosian (1983): Estimating bioconcentration potential from octanol/water partition coefficients. In: D. Mackay, S. Paterson and St. J. Eisenreich (ed.): Physical Behaviour of PCBs in the Great Lakes, Ann Arbor Science, Ann Arbor, Michigan (USA), pp. 269 - 282.

Yen, C.P.C., Perenich, T.A. and G. L. Baughman (1989): Fate of Dyes in Aquatic Systems II. Solubility and Octanol/Water Partition Coefficients of Disperse Dyes. *Environ. Toxicol. Chem.* 8, 981 - 986.

Testing Bioconcentration of Organic Chemicals with the Common Mussel (Mytilus edulis)

W. Ernst, S. Weigelt, H. Rosenthal and P.-D. Hansen

8.1 Introduction

Bioconcentration of persistent pollutants can result in levels of these compounds high enough to produce harm to exposed organisms. For the prediction of risks posed by elevated levels of pollutants in aquatic organisms it is necessary to determine the bioconcentration potentials of chemicals in appropriate tests. Enrichment of chemicals in organisms has been described in the literature as accumulation, bioaccumulation, bioconcentration and biomagnification. For clarity, in this paper only the term bioconcentration is used defined as the direct uptake of a substance by an organism from water without consideration of the ingestion of contaminated materials and resulting in substance levels in the organism beyond that in the ambient water. There are two modes of testing under standardized conditions: the static and semi-static exposure test and the flow through test. The results of these tests are reported as the bioconcentration factor (BCF); i.e. the quotient of the concentration of the substance under study in the organism and the ambient water. The BCF can be calculated on a fresh weight or lipid weight basis and can be determined at any time during the exposure period, but values derived at conditions reflecting steady state should be used for predictions and for comparative purposes.

Bioconcentration tests with fish are described as static, semi-static and flow-through tests in the OECD-Guidelines for testing chemicals. Previous work on bioconcentration of organic chemicals in the common mussel, *Mytilus edulis* were promising in using mussels as test organisms for various reasons, such as availability, high filtration rate, appropriate size, easy maintenance and low fat content (Ernst 1977).

Up to now it was largely unknown whether mussels can be used in a short term bioaccumulation test where one of the criteria to be met was year round availability of test animals exhibiting similar physiological condition. It was assumed that for such short term tests adaptation to experimental conditions may be a minor source of variability influencing the results. Furthermore it was postulated that the body burden with potentially harmful contaminants in mussels originating from commercial raft cultures of various geographical regions may be low, showing little fluctuations with time and space. Whether wild or cultured mussels can be more effectively used as experimentals in a standardised accumulation test is also an unanswered question. However, it is well established that blue mussels cultured in various parts of the world can be produced with consistent quality in terms of meat content, shell form and thickness and individual size. It will largely depend on the purpose of their use whether the encountered variations in initial quality are sufficiently small to be tolerated as variable in any of the samples to be used experimentally. Mussels of a particular size have extensively been used in the international "mussel watch" programme, where they are exposed to various water bodies as a tool in standard monitoring procedures, (Goldberg et al. 1978, Goldberg 1980). The question whether mussels from different origins at different times of the year will meet these requirements for a standardized bioaccumulation test is also addressed in this study.

As an alternative test or as a test additional to the OECD-static fish test, the mussel test was elaborated using test substances with different log P-values and under various maintenance regimes (Ernst et al. 1987).

8.2 Elaboration of the Test

8.2.1 General Outline

For reasons of practicability the working schedule was constituted of essentially three parts, according to Figure 8.1:

a) the supply and maintenance of mussels, operated by the Biologische Anstalt Helgoland (BAH), Hamburg;

b) the investigation of stress on mussels posed by transport and maintenance, carried out at the Bundesgesundheitsamt (BGA), Berlin;

c) the bioconcentration test, testing the influence of various parameters on the bioconcentration factor, executed at the Alfred-Wegener-Institut für Polar- und Meeresforschung, Bremerhaven.

In this way the supply of test animals was guaranteed for the duration of the study from July 1983 to May 1985.

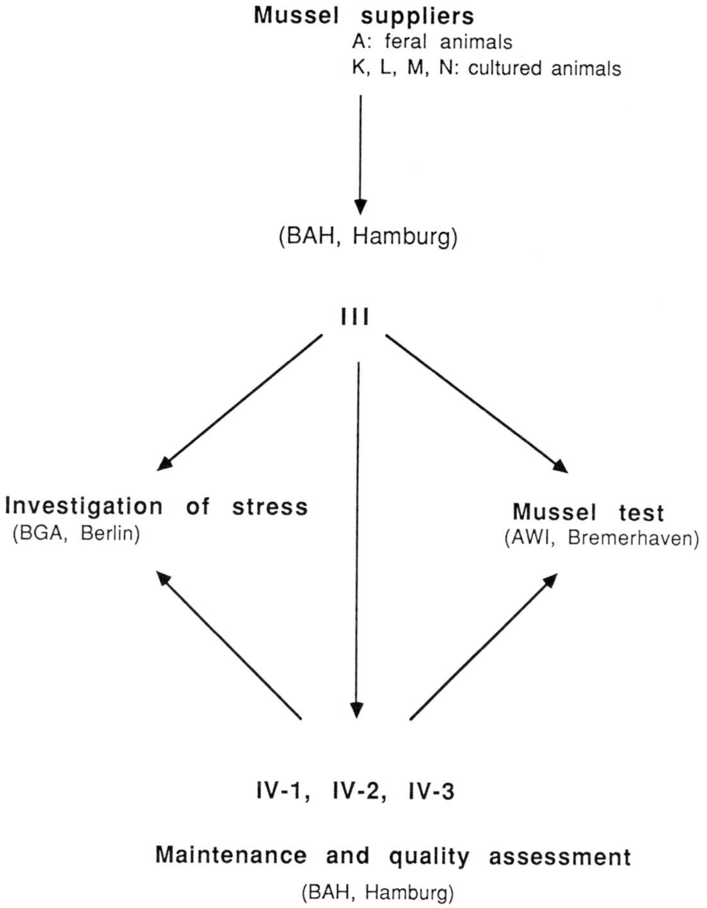

Figure 8.1 Schematic diagram for distribution and maintenance of mussels

Sources:

Sources:	A	List, Sylt (FRG)
	K	Yerseke (Holland)
	L	Austevall (Norway)
	M	Vigo (Spain)
	N	Göteborg (Sweden)

III: direct shipment from Hamburg without intermediate maintenance
IV-1, IV-2, IV-3: 2,-4,-8-weeks maintenance
Group IV-3 was fed at different levels, see Figure 8.6 and Table 8.2 for symbols

8.2.2 Mussels

8.2.2.1 Supply, Transport and Maintenance

In order to use physiologically comparable test animals at different seasons, it was planned to obtain the mussels from Spain, Portugal, France, Holland, Denmark, Sweden, Norway and the FRG. Due to difficulties with customs, lacking interest of mussel farmers, and embargo on exports when mussels became toxic by "red tides", preferably animals from Norway and the FRG were used.

The samples from Sylt (FRG) were taken from continuously submerged beds (Lister Tief) and the intertidal zone (Königshafen).

For the shipment of mussels parcels have been proven useful when isolated with styrofoam plates, 5 cm. Before shipment, 2-3 kg of animals were packed wet in plastic bags. According to outside temperature 1-3 cooling elements were added. For temperature deviation during transport see Table 8.1. Mussels used from Austevall (Norway) were from raft cultures, whereas those from List (FRG) were feral catches. Mussels from raft cultures should quickly be shipped, since in a number of cases the shells were slightly opened.

8.2.2.2 Quality of Mussels

After receipt from the suppliers the mussels were either used in the test after a 2-3 days adaptation period or maintained up to 8 weeks without food; some lots were also fed with algae (Table 8.2).

Also the background contamination by chlorinated hydrocarbons was determined. Compared to levels of the test compounds after exposure, the original contamination was found to be negligible (Table 8.3).

As further quality parameters, the shell length, soft tissue and fat content was determined, as well as the condition factor, defined as

$$K = G/L^3 \times 100, \text{ where}$$

G = soft tissue, wet or dry weight (g)
L = shell length (cm)

Data from all deliveries of mussels are shown in Table 8.4. The seasonal variations of physiological data are shown in Figures 8.2 and 8.3.

8.2.3 The Test Procedure and Variation of Experimental Parameters - Principle of the Test

In the proposed static test mussels are exposed to the substance dissolved in sea water at a suitable concentration. The decrease of the substance concentration due to uptake by the animals is determined at given intervals of time until the concentration remains practically constant over a defined period. This condition is defined as steady state.

Exposed organisms are analysed after termination of the exposure and accurate analysis of exposure water is carried out at the same time. Substance concentrations must be low enough to avoid detrimental effects to the mussels but high enough to allow chemical analysis of exposure water and organisms.

A detailed description of the test procedure is presented as Guideline draft in this paper. A flow scheme is outlined in Figure 8.4, and typical curves for the decrease of substance concentrations in exposure water within one week is shown in Figure 8.5.

8.2.4 Testing the Effects of Various Parameters on the Bioconcentration Factor of γ-HCH, Dieldrin and pp'-DDD

The bioconcentration factor as measured in a specified test can be influenced by a number of factors, such as origin of test animals, type of breeding, e.g. feral or cultured, maintenance regimes, physiological conditions, substance concentrations and others. In a number of exposure experiments the effect of these factors has been tested.

8.2.5 Origin and Maintenance

Test results with mussels from different countries and maintained under various feeding regimes are shown in Figure 8.6, exemplary from some exposure experiments with γ-HCH and pp'-DDD; all results from up to 82 exposure experiments were combined for γ-HCH, dieldrin and pp'-DDD for mussels from the two almost entirely used locations in Figure 8.7. Significant differences of BCF's with mussels from these two locations could not be observed; also different times of maintenance with and without feeding the animals were not likely to influence the BCF's severely.

Table 8.1 Variation of temperature during transport of mussels
a: packing before shipment, b: arrival, c: average deviation,
n: number of observations
A: destination Berlin, parcel service
B: destination Berlin, parcel service
C: destination Bremerhaven, parcel service
D: destination Hamburg, air freight
n: number of deliveries

Date	n	Duration of transport (h)	Temperature (°C) (a)	(b)	(c)
A Aug.-Oct. 83	7	~ 8 - 12	14.2 ± 1.2	13.1 ± 4.3	3.0 ± 2.3
B Oct. 83-Oct. 85	50	26.1 ± 1.5	13.1 ± 2.0	11.7 ± 3.7	3.5 ± 2.6
Winter 83/84 (Oct.-April)	14	26.1 ± 1.2	13.5 ± 1.3	10.1 ± 2.0	-3.4 ± 2.1
Summer 84 (May-Sept.)	14	25.3 ± 2.2	13.1 ± 2.3	14.3 ± 3.1	1.2 ± 3.3
Winter 84/85 (Oct.-April)	33	26.5 ± 1.1	12.9 ± 1.9	10.6 ± 3.2	-2.3 ± 3.6
C Oct. 83-Oct. 85	67	~20	13.3 ± 1.8	10.8 ± 3.6	3.9 ± 2.6
Winter 83/84 (Oct.-April)	22	~20	13.1 ± 2.1	8.9 ± 3.1	-4.1 ± 4.2
Summer 84 (May-Sept.)	20	~20	13.8 ± 0.9	13.9 ± 2.8	0.4 ± 2.9
Winter 84/85 (Oct.-May)	36	~20	13.0 ± 2.1	10.0 ± 2.5	-3.0 ± 3.5
D total	9	12 - 14	10.2 ± 4.3	11.8 ± 4.3	3.1 ± 2.0

Table 8.2 Average biomass and biomass loss of maintained mussels
Feeding regime:
*: once per week, **: twice per week, Δ: daily
others: without feeding, Average values, n = 50

Series No	Shell length (cm)	Average weight of soft tissue (g. dry weight) at delivery	after 4 weeks	loss of weight (%)	after 8 weeks	loss of weight (%)
1	3.58 - 3.63	0.43	0.38	11.63	0.35	18.60
2	4.37 - 4.42	0.62	0.57	8.06	0.51	17.74
3	2.17 - 2.22	0.03	0.02	33.33	0.02	33.33
4	2.03 - 2.07	0.03	0.03	0.00	0.02	33.33
					0.02*	33.33
5	3.79 - 3.82	0.22	0.17	22.73	0.16*	27.27
6	3.36 - 3.46	0.08	0.05	37.50	0.05*	37.50
7	2.68 - 2.76	0.06	0.06	0.00	0.05	16.67
					0.05*	16.67
8	4.71 - 4.78	0.40	0.35	12.50	0.33	17.50
					0.32*	20.00
					0.34**	15.00
9	4.03 - 4.09	1.63	1.53	6.13	1.44	11.66
					1.73*	+6.13
					1.58**	3.07
10	4.28 - 4.32	0.38	0.39	+2.56	0.32**	15.79
11	5.44 - 5.50	1.05	0.95	9.52	0.85**	19.05
12	5.60 - 5.66	1.15	1.13	1.74	1.05 Δ	8.70
13	4.37 - 4.60	0.44	0.54	+18.52	0.44 Δ	0.00
14	5.77 - 5.85	1.46	1.40	4.11	1.22**	16.44
15	4.49 - 4.70	0.64	0.63	1.56	0.60**	6.25
16	5.19 - 5.28	1.05	0.98	6.67	0.85**	19.05
17	4.88 - 4.90	1.35	1.12	17.04	0.90**	33.33
18	5.48 - 5.52	1.04	0.99	4.81	0.88**	15.38
19	4.56 - 4.76	0.78	0.79	1.27	0.62**	20.51
20	6.03 - 6.19	1.58	1.22	22.78	1.13**	28.48
21	3.78 - 3.94	0.27	0.28	+3.57	0.24**	11.11
22	3.36 - 3.43	0.17	0.15	11.76	0.11**	35.29
23	4.95 - 5.03	0.69	0.59	14.49	0.48**	30.43

Table 8.3 Contamination of mussels by organohalogen compounds at delivery without maintance; concentrations in ng/g fresh weight
No.1, 2, 8, 11, 12, 14, 16, 18, 20, 23, 25, 26, 29, 31 from List (FRG)
No.3 (Holland); No. 6 (Spain); No. 7 (Sweden)
others: Norway; Σ DDT: DDT + DDD + DDE

No	Date	PCB	α-HCH	γ-HCH	Dieldrin	Σ DDT	HCB
1	07.07.83	25.7	1.4	5.3	0.7	1.8	0.22
2	26.08.83	30.6	-	4.8	4.6	2.9	-
3	30.09.83	-	-	-	-	-	-
4	15.11.83	-	-	9.1	-	-	-
5	17.01.84	6.0	-	4.0	-	< 1.0	-
6	03.02.84	8.9	-	0.8	-	< 1.0	0.06
7	10.02.84	11.5	-	1.8	-	1.6	0.07
8	30.03.84	16.4	0.5	2.5	1.1	2.9	0.07
9	11.04.84	2.0	0.6	1.9	1.1	< 1.0	0.03
10	26.06.84	13.7	2.7	3.6	2.4	11.9	0.15
11	13.07.84	18.0	1.0	1.1	0.7	4.3	0.07
12	03.08.84	6.6	0.6	1.5	0.6	2.6	0.03
13	24.08.84	3.4	0.4	4.1	0.8	< 1.0	-
14	14.09.84	17.2	3.3	3.8	0.7	3.7	0.04
15	28.09.84	4.2	1.8	4.6	0.7	2.4	0.04
16	12.10.84	24.9	0.5	1.0	1.0	2.5	-
17	23.10.84	6.2	1.1	4.2	0.6	2.8	0.07
18	12.11.84	8.8	-	2.1	0.7	< 1.0	-
19	04.12.84	4.6	0.6	1.7	0.7	1.5	-
20	13.12.84	24.0	0.8	2.9	4.0	2.9	0.12
21	16.01.85	4.4	0.4	1.6	1.5	3.2	0.07
22	15.03.85	5.7	0.6	1.2	-	1.6	0.07
23	26.03.85	26.4	-	1.3	-	1.7	0.22
24	01.07.85	2.9	0.7	1.6	0.9	1.1	-
25	23.07.85	20.3	1.3	5.2	-	3.3	0.08
26	06.08.85	13.1	0.8	2.3	1.0	1.9	0.08
27	19.08.85	4.5	0.7	3.9	0.6	1.2	0.03
28	02.09.85	2.1	1.0	2.8	0.3	1.1	-
29	16.09.85	16.8	0.7	7.1	0.6	2.4	0.08
30	27.09.85	7.8	1.4	3.6	0.8	1.3	0.07
31	18.10.85	19.3	0.5	2.4	n.a.	1.2	0.10

Table 8.4 Shell lenght, soft tissue wet weight, lipid content and condition factor of mussels used in the test
EOM: Extractable organic matter
For numbers and symbols see Table 8.3 and Figure 8.1

No		length (mm)	soft tissue (g)	lipid (EOM) (%)	condition factor (fresh weight)
1	I A/III	36.6 ± 2.9	1.8	0.8 ± 0.2	3.7
	IV-1	37.3 ± 3.0	1.8 ± 0.4	0.9 ± 0.2	3.5
	IV-2	36.9 ± 2.7	1.8 ± 0.4	1.0 ± 0.3	3.6
	IV-3	36.9 ± 3.0	1.8 ± 0.5	0.9 ± 0.1	3.6
2	I A/III	44.5 ± 2.2	2.6 ± 0.4	1.0 ± 0.1	3.0
	IV-1	44.6 ± 2.1	2.8 ± 0.4	1.1 ± 0.1	3.2
	IV-2	44.7 ± 2.6	2.8 ± 0.5	1.0 ± 0.1	3.1
	IV-3	44.6 ± 2.2	2.4 ± 0.3	1.1 ± 0.1	2.7
3	II K/III	22.6 ± 1.2	0.3 ± 0.1	0.4 ± 0.2	2.6
	IV-1	22.3 ± 1.5	0.2 ± 0.1	0.3 ± 0.1	1.8
	IV-2	21.7 ± 1.4	0.2 ± 0.1	0.2 ± 0.1	2.0
	IV-3	21.5 ± 1.2	0.2 ± 0.1	0.3 ± 0.1	2.0
4	II L/III	20.4 ± 2.4	0.2 ± 0.1	0.4 ± 0.1	2.4
	IV-1	20.4 ± 1.7	0.2 ± 0.1	0.4 ± 0.1	2.4
	IV-2	21.4 ± 2.8	0.3 ± 0.1	0.5 ± 0.2	3.1
	IV-3	22.3 ± 2.2	0.3 ± 0.1	0.6 ± 0.1	2.7
	*IV-3	22.3 ± 2.2	0.2 ± 0.1	0.4 ± 0.1	1.8
5	II L/III	42.4 ± 5.5	1.8 ± 0.7	0.7 ± 0.1	2.4
	IV-1	39.3 ± 5.6	1.5 ± 0.7	0.6 ± 0.1	2.5
	IV-2	44.1 ± 5.2	1.8 ± 0.5	0.7 ± 0.1	2.1
	IV-3	39.6 ± 5.9	1.3 ± 0.7	0.6 ± 0.1	2.1
	*IV-3	39.1 ± 6.0	1.2 ± 0.6	0.7 ± 0.1	2.0
6	II M/III	35.6 ± 3.4	0.7 ± 0.3	0.4 ± 0.1	1.6
	IV-1	33.2 ± 1.2	0.7 ± 0.3	0.4 ± 0.1	1.9
	IV-2	32.7 ± 4.6	0.6 ± 0.2	0.3 ± 0.1	1.7
	*IV-3	30.3 ± 10.2	0.5 ± 0.2	0.4 ± 0.0	1.8
7	II N/III	29.4 ± 2.7	0.7 ± 0.2	0.6 ± 0.1	2.8
	IV-1	26.2 ± 2.9	0.4 ± 0.1	0.4 ± 0.1	2.2
	IV-2	29.5 ± 2.9	0.5 ± 0.1	0.4 ± 0.1	1.9
	IV-3	28.3 ± 3.0	0.5 ± 0.1	0.4 ± 0.1	2.2
	*IV-3	28.8 ± 3.0	0.4 ± 0.1	0.4 ± 0.1	1.7

Table 8.4 continued

8	I A/III	46.9 ± 2.5	2.5 ± 0.5	0.9 ± 0.1	2.4
	IV-2	46.7 ± 2.5	2.2 ± 0.4	0.8 ± 0.1	2.2
	IV-3	46.3 ± 3.3	1.8 ± 0.6	0.7 ± 0.1	1.8
	*IV-3	47.7 ± 3.0	2.0 ± 0.5	0.8 ± 0.1	1.8
	**IV-3	46.5 ± 2.4	1.9 ± 0.5	0.8 ± 0.2	1.9
9	II L/III	41.1 ± 6.7	1.9 ± 0.9	1.1 ± 0.1	2.7
	IV-2	43.2 ± 6.2	2.1 ± 1.0	0.8 ± 0.1	2.6
	IV-3	41.6 ± 6.6	1.8 ± 1.0	0.7 ± 0.0	2.5
	*IV-3	44.8 ± 7.6	2.2 ± 1.0	0.7 ± 0.1	2.4
	**IV-3	43.4 ± 6.4	1.9 ± 0.8	0.7 ± 0.1	2.3
10	II L/III	44.0 ± 6.5	2.1 ± 1.0	1.3 ± 0.1	2.5
	IV-1	46.1 ± 5.6	2.4 ± 0.8	1.3 ± 0.3	2.4
	IV-2	43.1 ± 6.4	2.0 ± 0.9	0.9 ± 0.4	2.5
	***IV-3	45.3 ± 5.2	2.1 ± 0.8	1.1 ± 0.1	2.3
11	I A/III	53.5 ± 3.2	4.1 ± 0.9	1.3 ± 0.5	2.7
	IV-1	53.0 ± 4.4	4.2 ± 0.9	0.9 ± 0.2	2.8
	IV-2	54.5 ± 4.2	4.1 ± 0.9	1.0 ± 0.1	2.5
	***IV-3	55.9 ± 3.0	4.0 ± 0.7	1.0 ± 0.1	2.3
12	I A/III	54.7 ± 3.4	4.7 ± 0.7	1.2 ± 0.2	2.9
	IV-1	54.6 ± 3.1	4.6 ± 0.8	1.1 ± 0.1	2.8
	IV-2	56.5 ± 2.9	4.7 ± 0.8	1.3 ± 0.1	2.6
	****IV-3	56.4 ± 2.2	4.2 ± 0.7	1.1 ± 0.0	2.3
13	II L/III	48.9 ± 5.8	3.1 ± 1.0	1.1 ± 0.1	2.7
	IV-1	47.2 ± 7.1	2.9 ± 1.2	1.0 ± 0.1	2.8
	IV-2	46.3 ± 4.6	2.8 ± 0.9	1.1 ± 0.1	2.8
	****IV-3	49.7 ± 5.1	3.0 ± 0.9	1.0 ± 0.1	2.4
14	I A/III	58.1± 2.9	5.6 ± 0.9	1.3 ± 0.1	2.9
	IV-1	57.8 ± 2.7	5.3 ± 0.8	1.1 ± 0.0	2.7
	IV-2	57.5 ± 2.4	5.2 ± 0.7	1.1 ± 0.0	2.7
	***IV-3	58.0 ± 2.5	5.1 ± 0.8	1.2 ± 0.1	2.6
15	II L/III	47.5 ± 5.5	3.3 ± 1.1	1.5 ± 1.5	3.1
	IV-1	47.0 ± 4.9	3.1 ± 1.4	1.2 ± 0.1	3.0
	IV-2	44.7 ± 7.5	2.7 ± 1.2	1.4 ± 0.2	3.0
	***IV-3	48.9 ± 5.6	3.4 ± 1.2	1.5 ± 0.1	2.9
16	I A/III	52.0 ± 0.0	4.1 ± 0.7	1.3 ± 0.0	2.9
	IV-1	52.0 ± 0.0	4.1 ± 0.0	1.3 ± 0.3	2.9
	IV-2	52.7 ± 3.2	4.1 ± 0.7	1.2 ± 0.1	2.8
	IV-3	51.2 ± 2.9	3.5 ± 0.6	1.2 ± 0.2	2.6
17	II L/III	48.4 ± 5.7	4.8 ± 1.3	1.7 ± 0.1	4.2
	IV-1	50.0 ± 6.9	4.7 ± 1.4	1.9 ± 0.2	3.8
	IV-2	51.0 ± 7.6	5.2 ± 1.4	2.2 ± 0.2	3.9
	***IV-3	49.6 ± 6.4	4.9 ± 1.7	1.2 ± 0.1	4.0

Table 8.4 continued

18	I A/III	54.8 ± 2.6	4.2 ± 0.7	1.1 ± 0.0	2.6
	IV-1	56.8 ± 2.9	4.2 ± 1.5	1.2 ± 0.0	2.3
	IV-2	55.6 ± 2.6	4.1 ± 0.6	1.1 ± 0.0	2.4
	***IV-3	56.7 ± 4.2	4.1 ± 0.9	1.1 ± 0.2	2.2
19	II L/III	47.8 ± 6.5	3.8 ± 2.0	1.4 ± 0.2	3.5
	IV-1	49.7 ± 7.6	4.3 ± 1.8	1.1 ± 0.2	3.5
	***IV-3	47.0 ± 6.0	3.3 ± 1.4	1.1 ± 0.2	3.2
20	I A/III	63.0 ± 2.7	6.3 ± 1.1	1.1 ± 0.2	2.5
	IV-2	61.5 ± 2.9	5.6 ± 0.8	1.5 ± 0.1	2.4
	IV-3	61.9 ± 3.1	5.7 ± 0.9	1.2 ± 0.2	2.4
21	II L/III	41.5 ± 3.6	2.0 ± 0.6	1.2 ± 0.2	2.8
	IV-1	40.5 ± 4.7	1.7 ± 0.7	1.2 ± 0.2	2.6
	IV-2	37.6 ± 4.5	1.4 ± 0.6	1.1 ± 0.1	2.6
	IV-3	39.7 ± 4.9	1.4 ± 0.6	1.0 ± 0.1	2.2
22	II L/III	34.6 ± 5.3	1.1 ± 0.6	0.6 ± 0.1	2.7
	IV-1	34.0 ± 4.7	1.0 ± 0.5	0.6 ± 0.1	2.5
	IV-2	33.3 ± 4.3	0.8 ± 0.3	0.6 ± 0.1	2.2
	IV-3	32.0 ± 2.5	0.6 ± 0.2	0.5 ± 0.1	1.8
23	I A/III	50.6 ± 2.5	3.7 ± 0.8	1.0 ± 0.0	2.9
	IV-1	49.8 ± 2.5	3.0 ± 0.6	1.0 ± 0.1	2.4
	IV-2	49.8 ± 2.6	2.8 ± 0.5	1.0 ± 0.1	2.3
	IV-3	49.9 ± 2.6	2.6 ± 0.4	0.8 ± 0.1	2.1
24	II L/III	48.6 ± 4.3	3.9 ± 1.1	1.4 ± 0.2	3.4
25	I A/III	52.8 ± 1.7	5.3 ± 0.7	1.6 ± 0.2	3.6
26	I A/III	52.2 ± 1.5	5.3 ± 0.9	1.5 ± 0.1	3.7
27	II L/III	55.9 ± 3.0	5.8 ± 1.1	1.6 ± 0.2	3.3
28	II l/III	52.9 ± 2.2	5.1 ± 1.0	1.5 ± 0.2	3.4
29	I A/III	53.2 ± 1.6	5.5 ± 1.1	1.4 ± 0.2	3.7
30	II L/III	49.9 ± 4.8	4.7 ± 1.5	2.2 ± 0.2	3.8
31	I A/III	49.0 ± 7.3	5.1 ± 0.9	1.5 ± 0.1	4.3

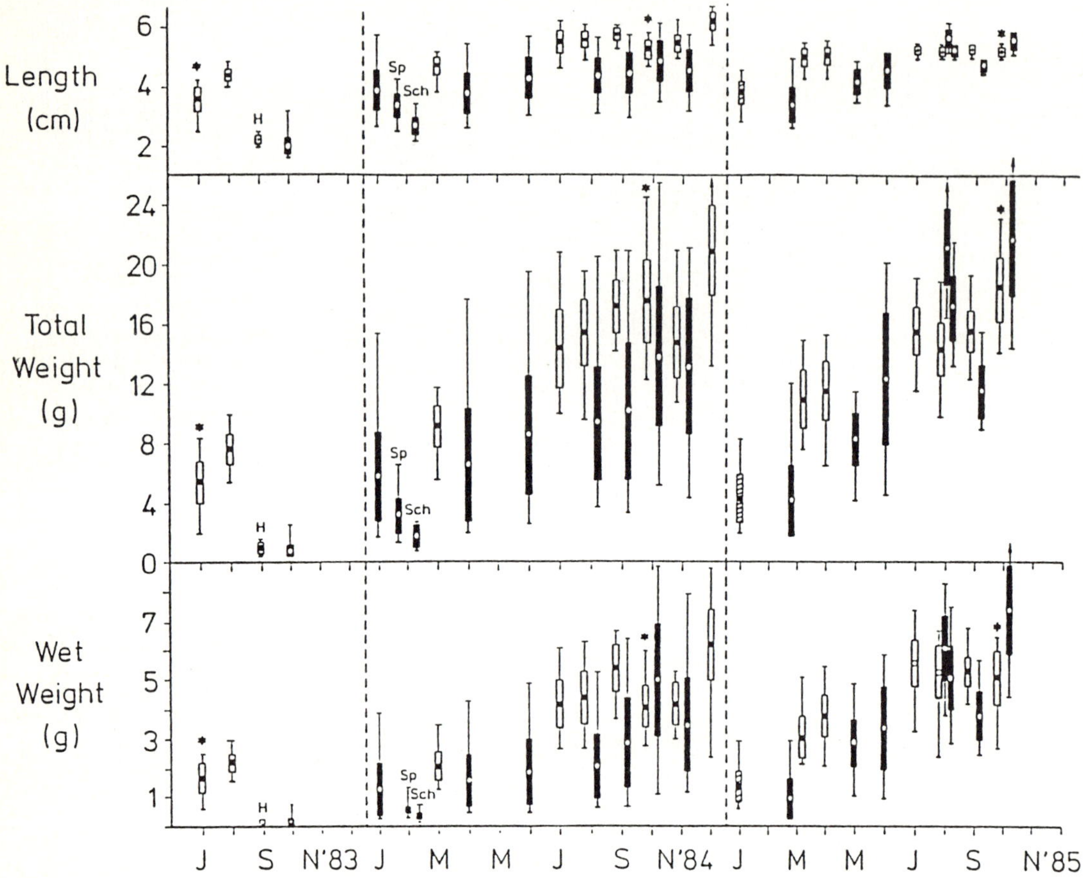

Figure 8.2 Characterization of seasonal differences in mussel samples of different origin. Data represent means, standard deviation (bars) and ranges (lines) for shell lenght, total weight and wet weight of soft tissue.
open bars = samples from List/Sylt (FRG)
black bars = samples from Norway
Sp = Spain; SCH = Sweden; H = Holland
Hatched bars = Norway, winter station
* = Intertidal samples

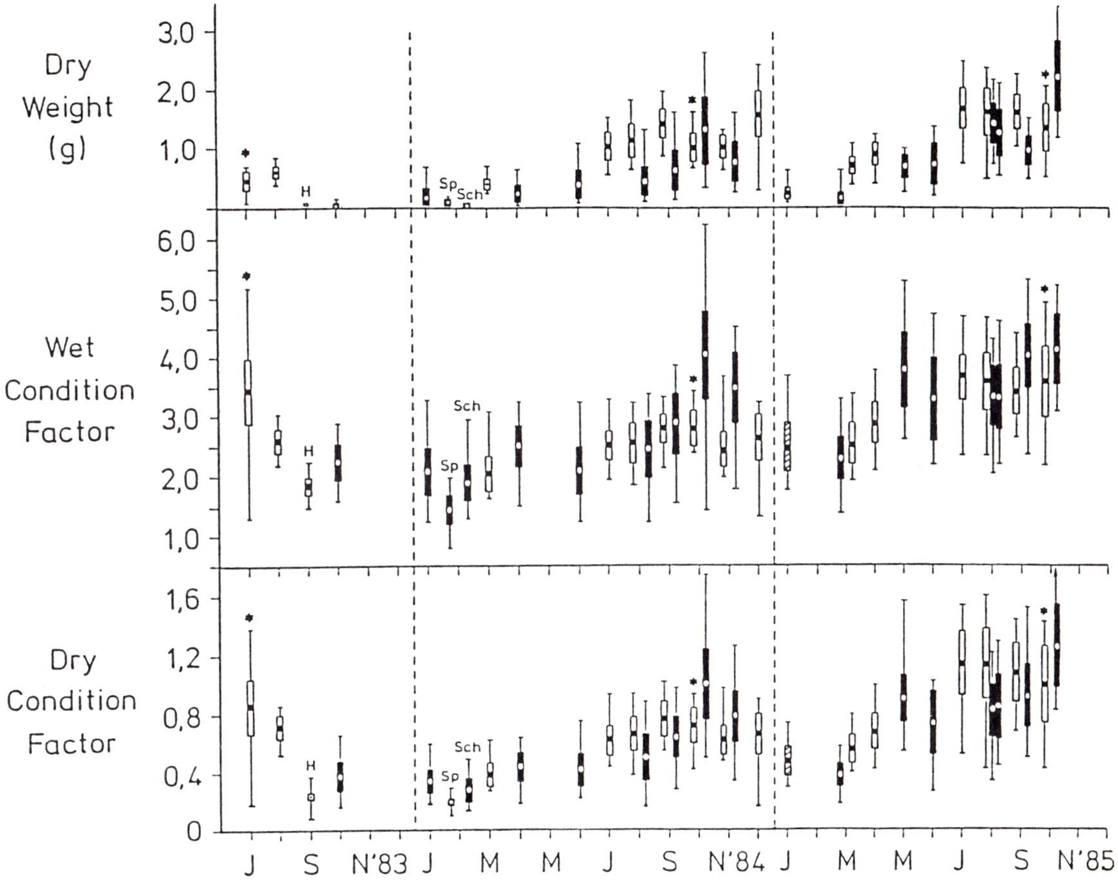

Figure 8.3 Saisonal variation of dry weight, wet and dry condition factor of mussels from different sources.
For symbols see Figure 8.2

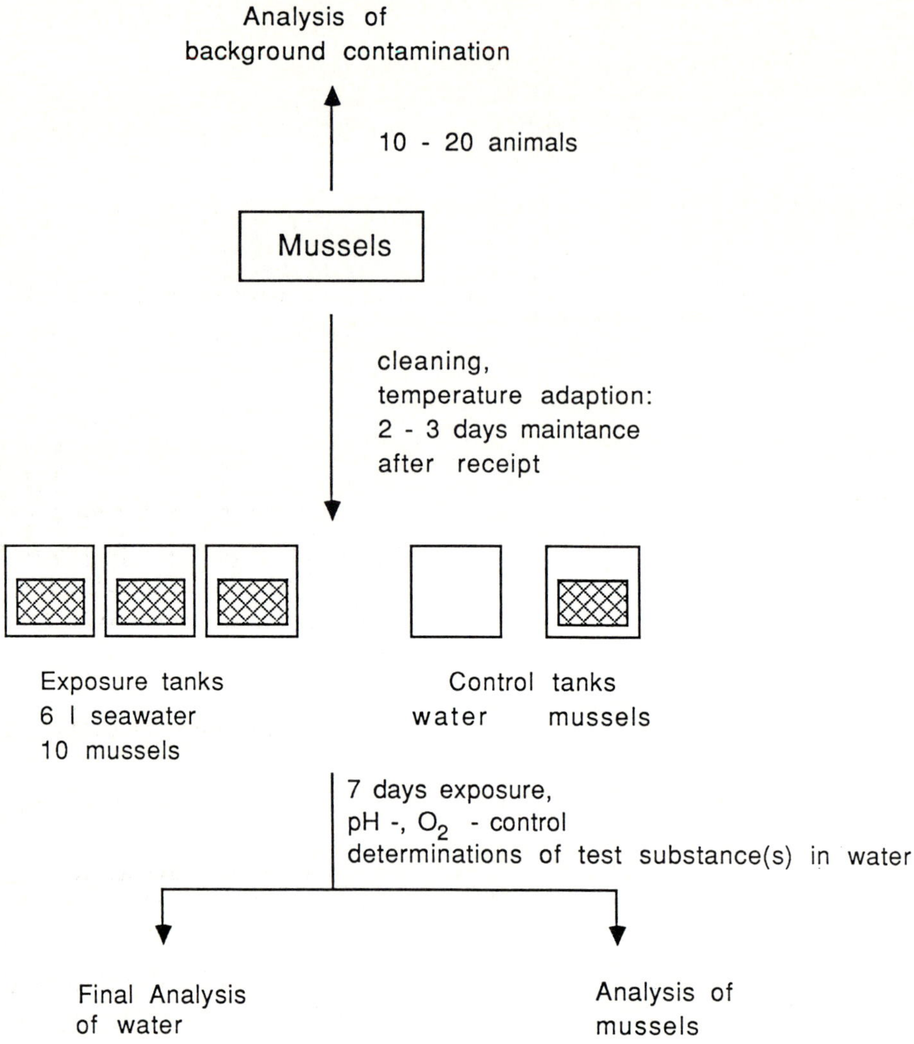

Figure 8.4 Schematic diagram, mussel test.

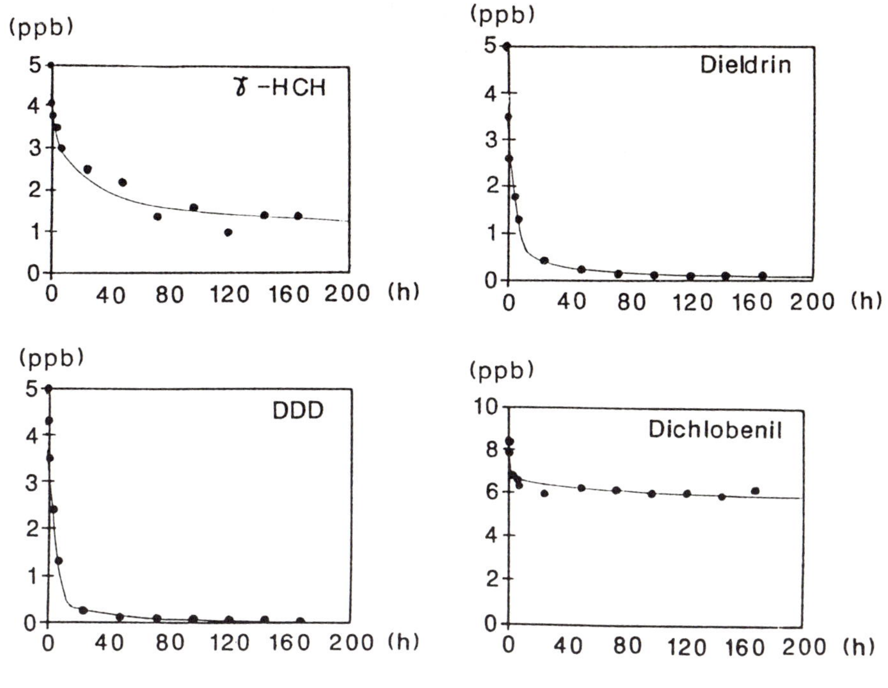

Figure 8.5 Decrease of substance concentrations in sea-water during exposure of mussels. γ-HCH, dieldrin and pp'-DDD at initial concentrations of 5 μg/l; dichlobenil at 10 μg/l.

8.2.6 Variation of Exposure Concentration

Most exposures were carried out at an initial substance concentration of 5 μg/l, but also concentrations of 2 and 10 μg/l were tested. The results for γ-HCH and pp'-DDD are listed in Table 8.5 indicating that in this range of substance concentrations differences of BCF are negligible.

8.2.7 Variation of the Biomass/Water-Relationship

A number of exposures were carried out at a lower biomass/water ratio than in the standard procedure, i.e. five instead of 10 mussels per 6 L of seawater. As shown in Table 8.6, lipid-related BCF's for γ-HCH were identical in both exposures, those for pp'-DDD were 25% higher in the low-biomass exposure.

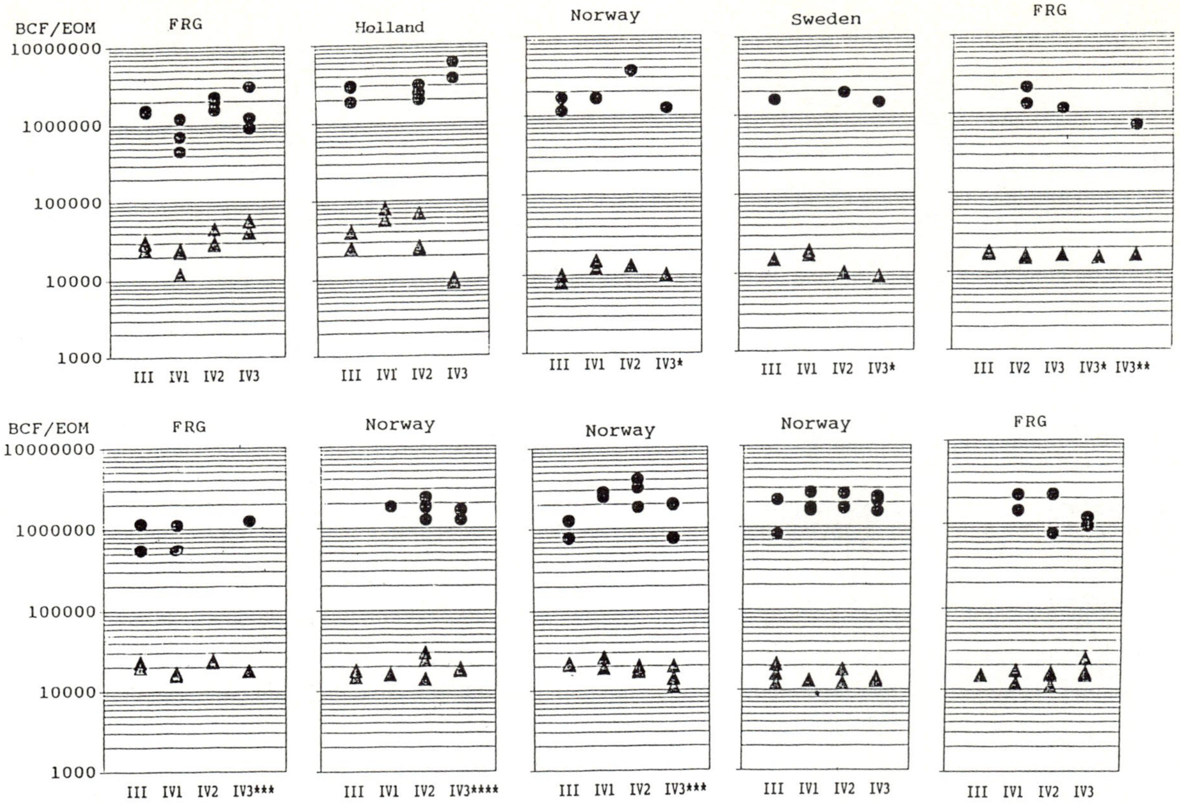

Figure 8.6 Bioconcentration factors for γ-HCH (▲) and pp'-DDD (●) based on lipid (EOM) exemplary for 10 exposure series with mussels from different countries and after different maintenance regimes (Log-scale).

III: without intermediate maintenance, IV-1: 2 weeks maintenance
IV-2: 4 weeks maintenance, IV-3: 8 weeks maintenance
*, **, ***, ****: 4 x, 16 x, 8 x and daily fed with algae

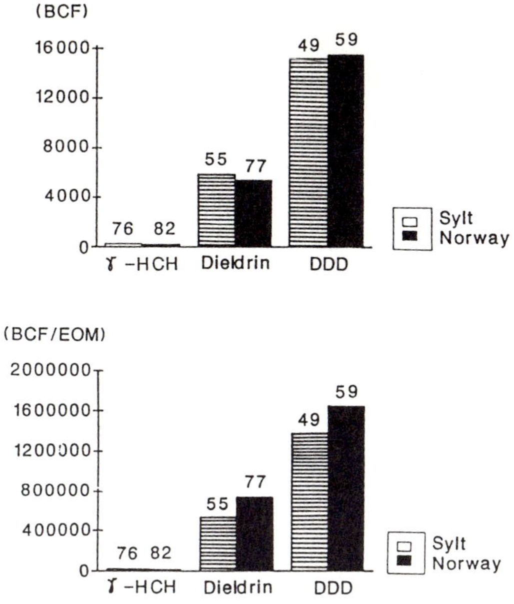

Figure 8.7 BCF's for γ-HCH, dieldrin and pp'-DDD for mussels from Norway (raft culture) and from List (Sylt)/FRG (feral animals). Average values from all experiments, related to fresh weight of soft tissues and to lipid (EOM). Numbers on bars: No. of experimental series.

Table 8.5 Bioconcentration factors for γ-HCH and pp'-DDD obtained at different exposure concentrations
BCF = related to fresh weight, BCF/EOM = related to lipid,
n = no of experimental series

γ-HCH	BCF	BCF/EOM	n
2 ppb	228	24000	32
5 ppb	199	19000	106
10 ppb	225	22000	47
DDD			
2 ppb	12627	1412000	15
5 ppb	14831	1596000	105
10 ppb	14572	1662000	34

Table 8.6 Bioconcentration factors for γ-HCH and pp'-DDD obtained at different biomass water-ratios
BCF = related to fresh weight, BCF/EOM = related to lipid

	10 Mussels/6 L Water		5 Mussels/6 L Water	
	BCF	BCF/EOM	BCF	BCF/EOM
γ-HCH	259	18000	238	17000
	114	7000	126	12000
	177	15000	120	11000
	61	14000	69	13000
	132	15000	151	16000
	269	24000	254	23000
	194	18000	181	20000
	139	15000	137	14000
	85	13000	49	13000
	159±72	15000±5000	147±69	15000±4000
DDD	27275	1922000	22858	2090000
	8789	762000	23434	2174000
	6882	1553000	10943	2089000
	10166	1175000	8840	922000
	14875	1325000	23025	2090000
	26592	2481000	31542	3725000
	14392	1546000	12847	1337000
	14135	2092000	6581	1709000
	15388±7680	1607000±546000	17509±8868	2017000±821000

8.2.8 Saisonality of the Lipid Content of Mussels

Lipid contents of mussels vary with season and have an influence on BCF's (Figures 8.8 and 8.9). For comparative purposes BCF values should therefore be related to the lipid content of the animals.

8.2.9 BCF of γ-HCH, Dieldrin and pp'-DDD in a Flow-Through Test

The title compounds were tested additionally in a flow-through test system. The results compiled in Figures 8.10 and Table 8.7 indicate that steady state conditions are attained for γ-HCH within one week, for dieldrin only after 30 days, but apparently not for pp'-DDD after 45 days of exposure. The BCF's obtained were in the same order of magnitude as those in the static test (Tables 8.7).

8.2.10 Comparison of Results obtained in the Static Mussel Test with those from other Tests

Average values of bioconcentration factors from all mussel experiments with γ-HCH, dieldrin and pp'-DDD are shown in Table 8.7. Additionally, the BCF's of dichlobenil, pentachlorophenol (PCP), p-chloroaniline and atrazin obtained in the static test are listed. Results from tests performed with fish according to OECD-tests 305 B and 305 C are also incorporated in Table 8.7 for comparison as far as the compounds in these tests are concerned.

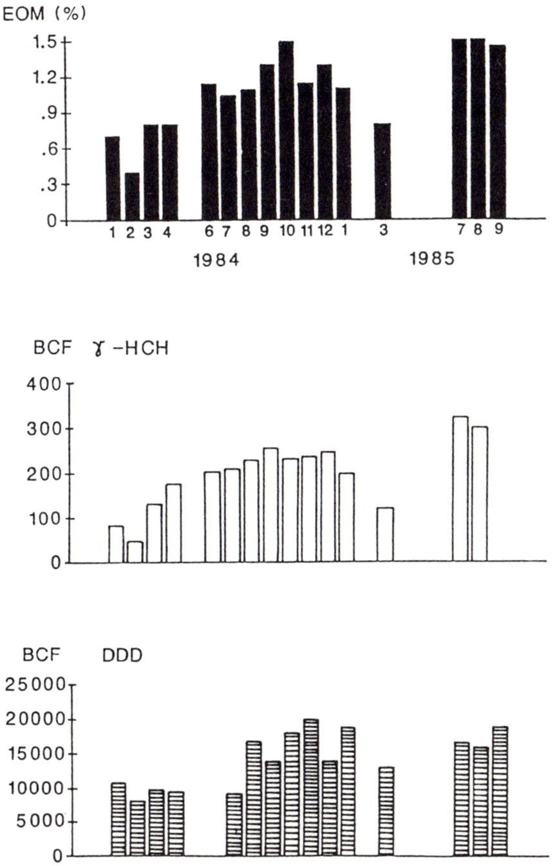

Figure 8.8 Saisonal variation of EOM in mussels and BCF's for γ-HCH and pp'-DDD related to fresh weight of soft tissue.

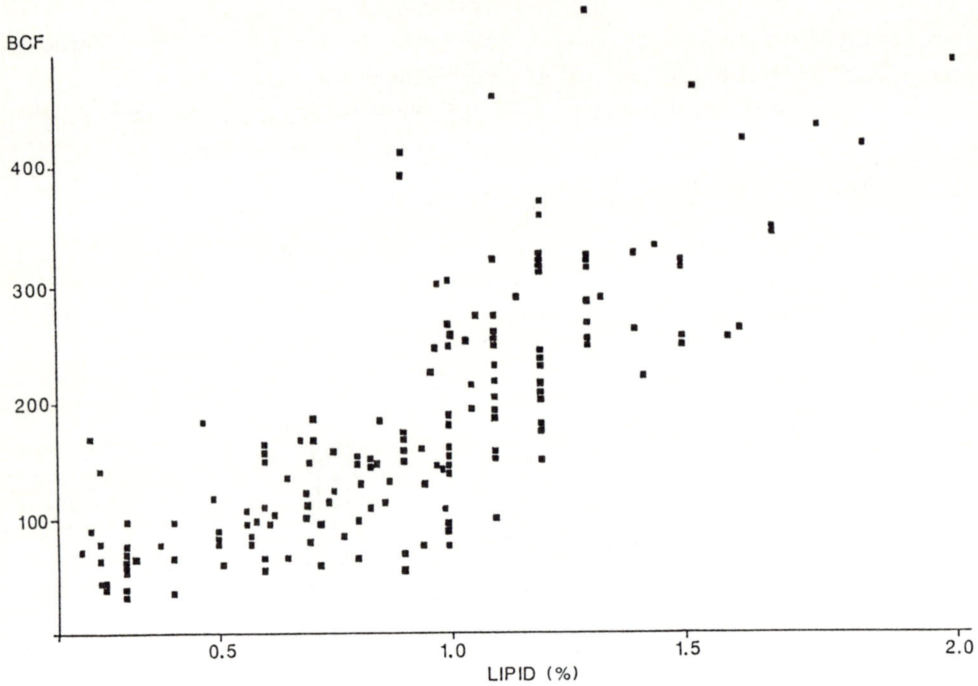

Figure 8.9 BCFs for γ-HCH versus lipid content.

8.2.11 Examination of Stress in Mussels caused by Transport and Maintenance

The blood cells (hemocytes) in the hemolymph of mussels are, among others, responsible for the elimination of invading microorganisms and foreign particles by phagocytosis. The process of phagocytosis may be characterized by the following steps: chemotactical activation of hemocytes, attachment of particles to the surface of the phagocytes, incorporation of the particles and their digestion.

As a measure of stress the phagocytosis of yeast cells was determined and expressed as index of phagocytosis (Figure 8.11). Mussels after transport and maintenance exhibited elevated phagocytosis activity (Figure 8.12) when compared with field measurements.

8.2.12 Metabolism

Mussels are capable of metabolizing xenobiotics to significant degrees and examples are given in Figures 8.13 - 8.15. In these cases, measured BCF's can be below those, calculated by quantitative structure-activity relationships (QSAR).

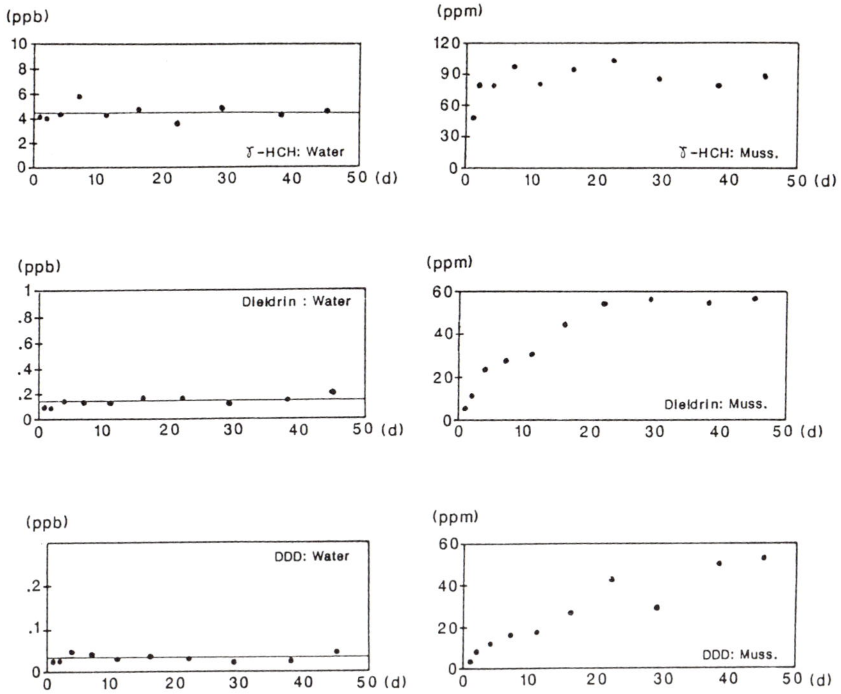

Figure 8.10 Uptake of γ-HCH, dieldrin and pp'-DDD in flow through tests. Left: Concentration of compounds indicated during exposure. Right: Increase of compound levels in mussels during exposure (values refer to lipid)

Table 8.7 Bioconcentration factors for different chemicals obtained with the mussel test in comparison to OECD-tests and a flow through test BCF, BCF (EOM): bioconcentration factors based on fresh weight and lipid basis OECD-test were performed using zebra fish, if not stated otherwise.

	Mussel test		OECD-test 305B	OECD-test 305C	
	BCF	BCF (EOM)	BCF	BCF	BCF (EOM) (calcd.)
γ-HCH	220 (205)	20000 (20000)	563[6] -601		16900 18000
Dieldrin	5170 (3750)	566000 (375000)			
DDD	14420 (12500)	1615000 (1250000)			
Dichlobenil	10	1000	43[4]		1300
PCP	170	19000	265[5] -314		8000 9400
p-Chloroaniline	90	15000	4[3] 7	17-20[7]	
Atrazin	n.d.	n.d.	6.8[1] 7.7[1] 3.5[2]		227 257

1) 10 μg/l, 100 μg/l; 2) 5000 μg/l, Golden orf; 3) 5000 μg/, 1000 μg/l; 4) 30 and 300 μg/l; 5) 0.4 μg/l, 4 μg/l; 6) 0.2 μg/l, 2 μg/l; 7) 250 μg/l

8.3 Summary

In this study the usefulness of test mussels originating from various sources has been tested and their initial quality been characterized in an attempt to standardize the procedures for a continuous provision of non-mature experimentals suitable for a bioaccumulation test. Because maturity is reached at different times of the year under different climatic conditions, it was originally intended to obtain shipments from Spain, Portugal, France, Netherlands, Denmark, Sweden, Norway and the Federal Republic of Germany. During the project, however, it was realized that a continous supply with mussels of similar initial quality could not be guarantied because of inconsistancies in long-distance transport conditions which could not always be controlled but would have substantial consequences for the bioconcentration test. Test animals were therefore obtain from Sylt/List (FRG) and from Norway.

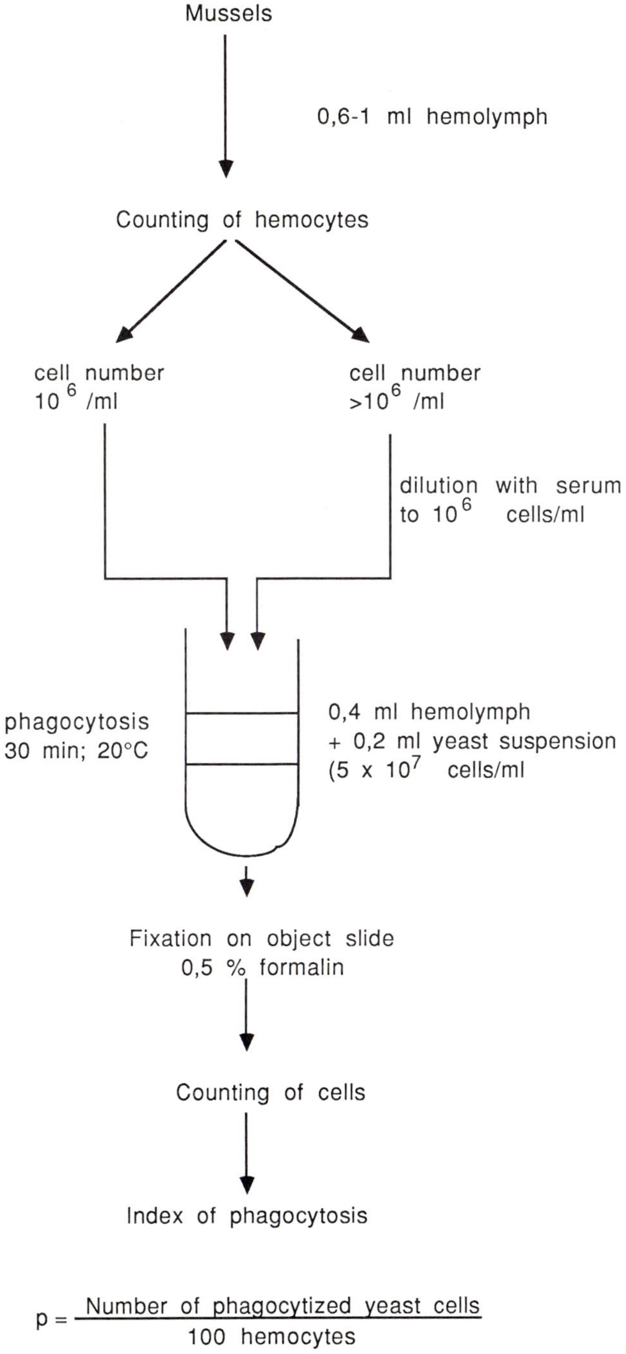

Figure 8.11 Schematic diagram of examination of stress in mussels.

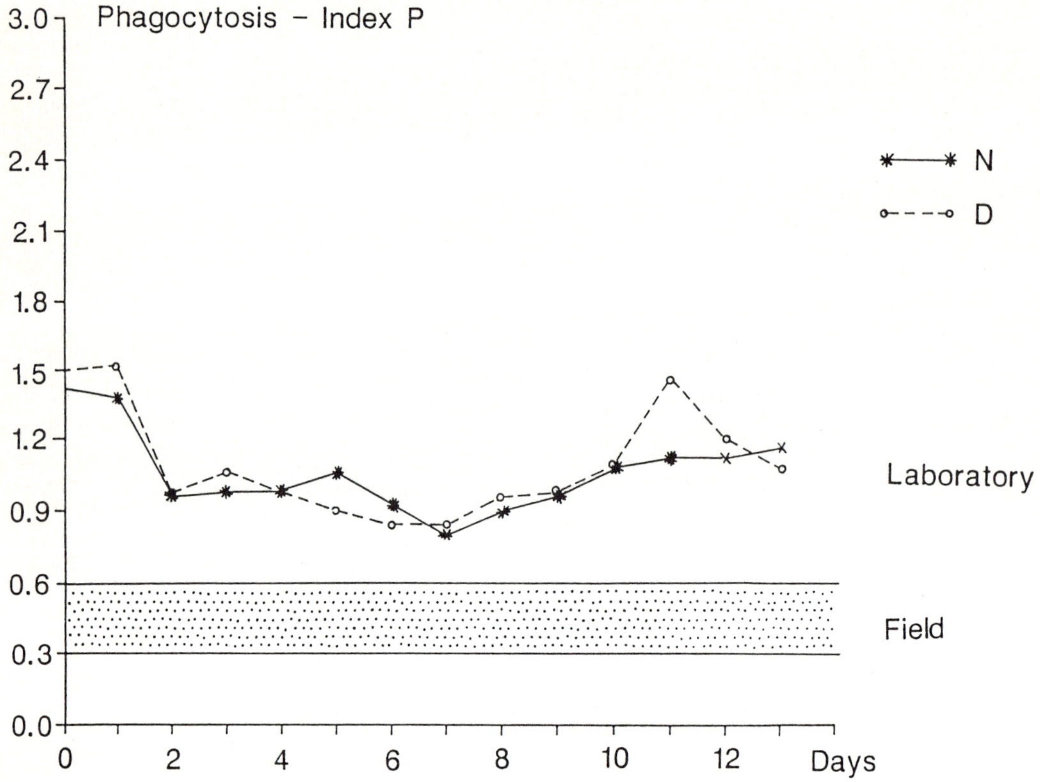

Figure 8.12 Average values of the index of phagacytosis as a measure for stress in mussels caused by transportation and maintenance in comparison with on site measurements.
N: mussels from raft cultures (Norway); D: mussels from List (FRG)

A bioconcentration test with mussels, Mytilus edulis,from the sources indicated, and as described in the proposed Guideline draft has been worked out and checked by varying a number of experimental parameters. Seven organic compounds, preferably γ-HCH, dieldrin and pp'-DDD exhibiting log p-values from about 4 to 6 were used in order to consider substances with different bioconcentration potentials and different physical properties. The test results were compared with those from a flow through test and from different OECD-tests.

Variation of test parameters, particularly different maintenance regimes are not likely to influence bioconcentration factors significantly. Comparison with other test procedures were satisfactory with values being in good agreement or at least in the same order of magnitude.

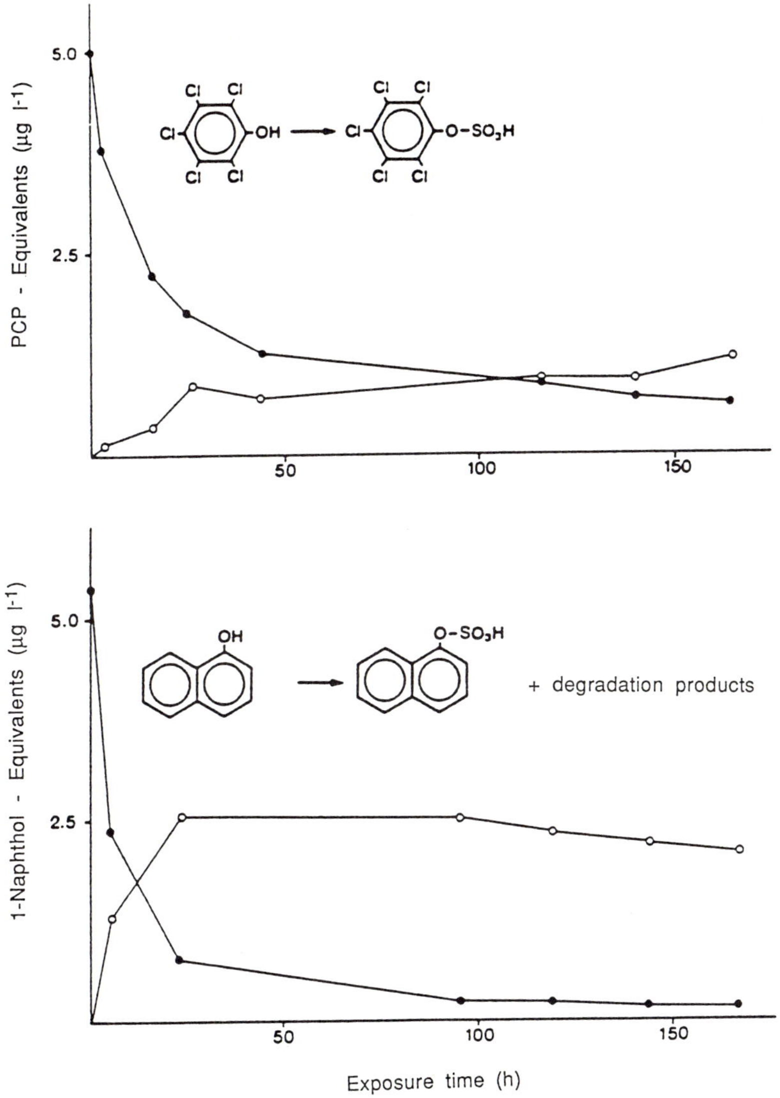

Figure 8.13 Biotransformation of 14C-pentachlorophenol and 14C-1-napththol in mussels (Mytilus edulis). Full dots: parent compound.

DBP

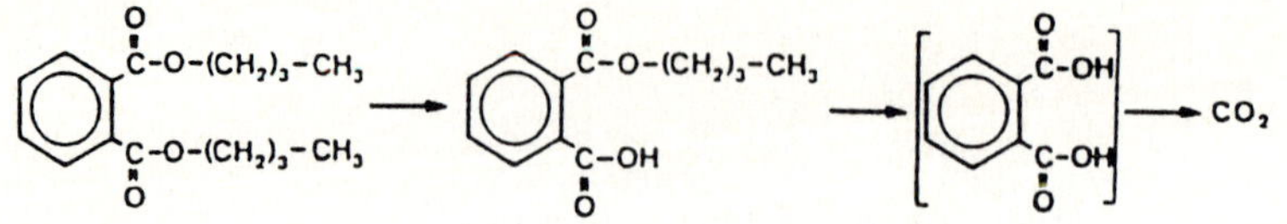

DEHP

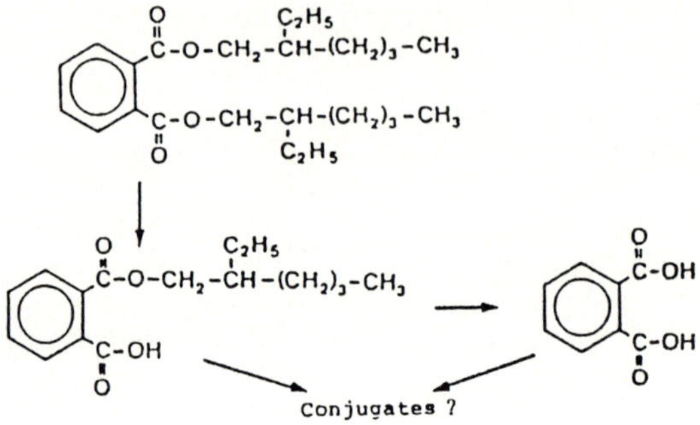

Figure 8.14 Biotransformation of 14C-dibutylphthalate (DBP) in mussels and diethylhexylphthalate (DEHP) in shrimps, oysters and fish

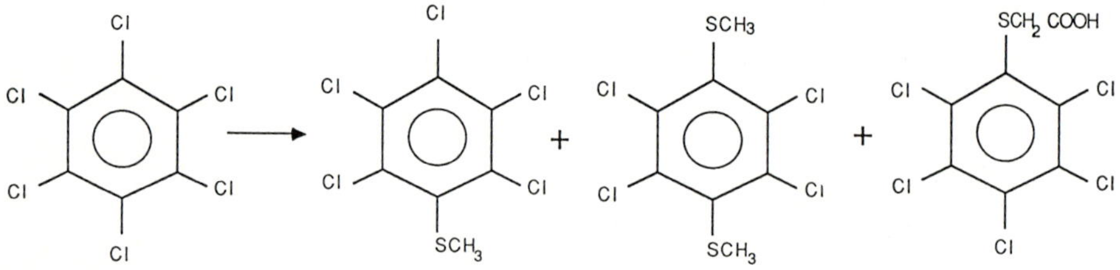

HCB

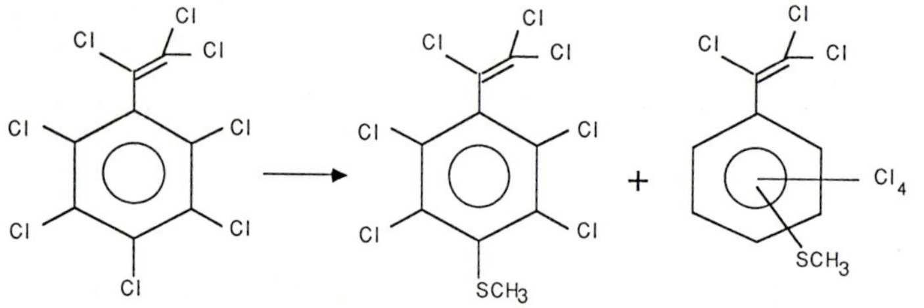

OCS

Figure 8.15 Biotransformation of 14C-hexachlorobenzene and octachlorostyrene in mussels (Mytilus edulis).

8.4 Guideline Draft: Bioaccumulation Testing with the Common Mussel, Mytilus edulis

8.4.1 Introductory Information

Prerequisites, data for test substances

- water solubility
- stability and reactivity in water
- acute toxicity to test organisms
- volatility
- analytical method specific to the test chemical

Qualifying statements

- This test procedure is applicable as long as the test chemical can be reliably analysed in the test organism and in the test medium.
- In case the test chemical is not stable and forms metabolites, the magnitude of the BCF must carefully be considered.

Recommendations

- Based on results from exposure experiments it is recommended that the BCF's should be related to the lipid content of the mussels.

8.4.2 Method

8.4.2.1 Introduction

The aim of bioconcentration tests is the determination of bioconcentration factors at steady state conditions. From appropriate experiments the common mussel, Mytilus edulis, was found to be suitable for the test: Mussels equilibrate very fast, they are abundant in coastal areas, easy to collect and are widely used as monitor organisms for environmental pollution.

The bioconcentration factor was determined by the quotient of the concentration of the compound in the test organism to that in water under steady-state conditions as defined in this paper.

8.4.2.2. Definition and Units

- Bioconcentration is involved where the concentration of the test substance in a specific tissue increases beyond that in the ambient water without taking into account the ingestion of materials contaminated with the test substance. The term "bioconcentration" will also be used when substances are adsorbed rather than absorbed by organisms.

- Bioconentration factor (BCF) is the ratio of the test substance concentration in the test mussel (c_M) to the concentration in the test water (c_W) at steady state.

- Steady state is a condition in which the amount of test material being taken up and depurated per unit of time is equal at a given substance concentration in water. Particularly in this paper it means that the concentration of the test compound in the static procedure is practically constant during a defined period.

8.4.2.3 Principle of the Test Method

The method is based on the static exposure of the mussels to different concentrations of the test compound in sea water. The bioconcentration process is followed by the analysis of water at defined time intervals until a constant concentration is attained, which is usually the case after 7 days. The mussels are then removed for analysis and also the concentration of the test compound in the water is determined very carefully. The BCF is then calculated by the quotient c_M/c_W, where c_M: level of the test compound in the mussel and c_W: concentration of the test compound in the water at the end of exposure.

8.4.2.4 Quality Criteria

The applicability of the test is limited, besides the toxicity of the compound, by various factors, such as volatility, stability in sea water, solubility and sensitivity of the analytical method.

Based on more than 100 experiments including up to 3 different concentrations of various test compounds a standard deviation of 15-25% for the bioconcentration factor was observed, qualifying the organisms by size, season and lipid content.

8.4.2.5 Description of the Test Procedure

Mussels can be obtained from various locations, including both natural and raft-cultured stocks. It must be guaranteed, however, that specimens are not mature and gonads are not in an advanced stage of development. For this reason it is recommended to use specimens below 4 to 5 cm total shell length.

A continuous and all-year-round supply of immature small mussels is not possible from one source only. Along many central and northern European coasts, Mytilicola infestations are heavy and render these stocks largely unsuitable for tests. In severe winters, supply might be impaired by heavy coastal ice formation. In other areas, toxic red tides prohibit harvest and export of mussels over extended time periods. Mussels from the List/Sylt Waddensea stocks and from Norwegian raft culture units turned out to provide the most reliable and practical sources, not affected by parasites and available for more than two thirds of the year. At least 3 locations, which are situated in different climates, should be selected for providing regular supplies, because of seasonal differences in maturation periods, annual cycle in condition factor and fat content, and because of possible shortages due to a number of logistic reasons.

8.4.2.6 Transportation

Depending on the season, transportation should take place in more or less isolated containers (styrofoam boxes; dimensions suitable for carrying at least 15-20 kg of mussels). It is further advisable to obtain mussels from locations which do not ask for more than 12 hours transportation time to the laboratory or holding facilities. This is mainly due to the reason that juvenile mussels are obtained from subtidal beds or from rafts and are not hardened to sustain long-term air exposure. Insulation should be sufficient to maintain temperatures slightly below or near test temperatures (8-10 °C). Temperatures in transport containers should be determined prior to departure from the collection sites and immediately after arrival. They should not deviate more than $\pm$ 3.0 °C.

8.4.2.7 Selection of Test Animals

Since mussels must be obtained from commercial sources under routine procedures, selection of suitable individuals cannot be undertaken on site and larger bulk shipments are required to allow large enough samples for later selection.

Upon arrival at the laboratory test specimens should be selected from the bulk shipment after carefully washing and cleaning the mussels from debris, accompanying fauna and flora and after removal of dead or open (weak) individuals. This procedure should be carried out in a cool room in order to maintain the animals within a reasonable temperature range. To avoid excessive handling time, initial selection should be done by size estimation rather than by exact measurement of shell lenght.

8.4.2.8 Live Storage for later Use

Mussels can be maintained without feeding at temperatures near or slightly below the test temperature for about two to four weeks without loss of the characteristical performance. Maintainence at lower temperatures is possible but not recommended, because of possible physiological adaptation responses. Stocking density in holding tanks should not exceed 100 kg/m3, preferably stocked in shallow tanks (water depth not greater than 25 cm). Water volume flowing through each tank should at least equal three times the tank volume per hour. Inlet arrangement should be installed in a way that allow to disperse the inflow water uniformly in the tank. Additional aeration is required in each tank. Water temperature should be maintained between 7 and 10 °C, pH values should not decrease below 7.8 at 30 ppt (‰) salinity. Holding tanks should be hooked up to a recirculation system employing a biological gravel/sand filter (grain size of coarse sand 2-5 mm diameter). Holding tanks should have a total volume not more than 33% of the total system volume. Biological filter must have minimum size of 100 L volume (filter bed, including voids) per 100 kg mussels. The biofilter should be backwashed in intervalls of about 6 weeks. Runoff water from holding tanks should pass a settling tank (retention time 45 min) before entering the biofilter. Settled sludge must be removed in regular intervals. Water losses through evaporation and leakage must be replaced by filtered fresh- and seawater, respectively. A 25% water replacement should be undertaken every 6 weeks, in connection with the biofilter backwash.

8.4.2.9 Analysis

The analytical procedures for the determination of the test compounds have to be applicable to water as well as to animals. Recovery has to be checked for the test compounds and blanks should be run for every batch. It is also useful to execute analysis of the "natural" level of contaminants in the test animals.

8.4.3 Performance of the Test

8.4.3.1 Exposure

According to Figure 8.16 an aquarium is installed for each concentration of the test com-

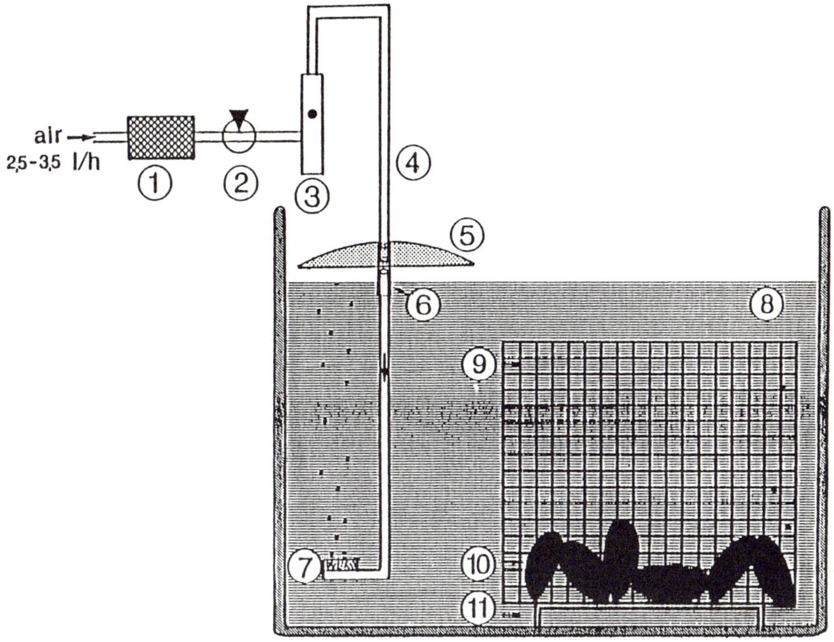

Figure 8.16 Aquarium arrangement for the test
1) charcoal filter; 2) needle valve; 3) flowmeter; 4) 3 mm glass tube;
5) watch glass (7 cm ϕ with 3.2 mm hole); 6) teflon tube; 7) glass frit;
8) 10 l glass aquarium containing 6 l sea water, temp. 10 $\pm$ 2 °C,
95-98 % O_2-saturation; 9) stainless steel basket with 1 cm mesh;
10) test animals; 11) Petridish

pound; additionally one blank without animals is used to control the loss of the test compound. The test compounds dissolved in ethanol or water are added to the aquaria and are well mixed with the sea water. Subsequently the initial concentration of the test compound is determined and the steel basket containing 10 animals is inserted. The oxygen saturation and the pH should be checked once every day but at least at the beginning and the end of the test (in these experiments it has been proven that O_2-saturation was practically constant at 95-98% and pH values exhibited variations of 0.1 unit). The concentration of the test substance in water is measured daily in order to construct an appropriate graph for the identification of the "steady state". At the end of the exposure animals are removed together with the steel basket and a water sample is drawn immediately for the exact determination of the final concentration (steady state concentration), c_W.

8.4.3.2 Preparation and Analysis of the Mussels

The shells of the test animals are opened according to Figure 8.17 by usig a scalpel, cutting the adductor muscle in the indicated direction. After a draining period of 3 min, in which the adherent water is removed, the soft tissue is dissected. The soft tissues might be analysed in single specimens: preferably samples of 5 animals according to the size of animals and to analytical procedures are combined for analysis. The analytical procedures used depend on the type and concentration of the substances. Recoveries should be checked for water and for animals.

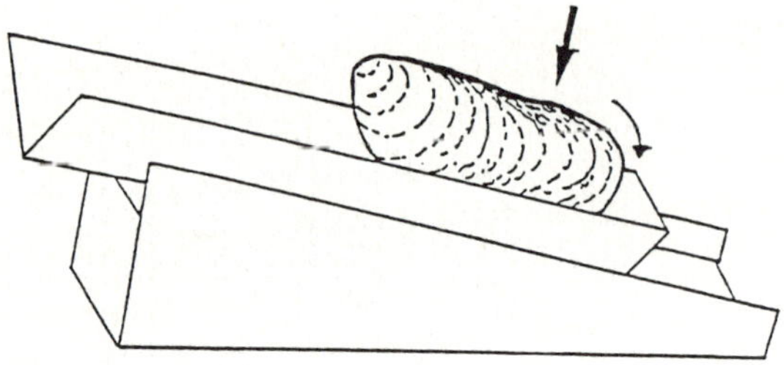

Figure 8.17 Preparation of soft tissue

8.4.3.3 Data and Reporting

The BCF is calculated as follows:

$BCF = c_M/c_W$ and $BCF_L = c_{ML}/c_W$ respectively.

BCF, BCF_L: Bioconcentration factor related to soft tissue fresh weight and to lipid content, resp.

c_M, c_{ML}: Concentration of substance in soft tissue on fresh weight and lipid basis (µg/g).

c_W: Concentration of substance in water when the test is terminated (µg/g).

8.5 References

Ernst, W. (1977). Determination of the bioconcentration potential of marine organisms - A Steady state approach. I. Bioconcentration data for seven chlorinated pesticides in mussels (Mytilus edulis) and their relation to solubility data. *Chemosphere* 6, 731-740

Ernst, W., Weigelt, S., Weigelt, V., Rosenthal, H. and Hansen, P.D.(1987). Entwicklung eines Bioakkumulationstests mit der Muschel *Mytilus edulis*. Forschungsbericht 106-020-24/05, Umweltbundesamt Berlin

Goldberg, E.D. (1980). The International mussel watch. Report of a workshop sponsored by the Environmental Studies Board Commission on National Resources. National Academy of Sciences, Washington, D.C., 248 pp.

Goldberg, E.D., Bowen, V.T., Farrington, J.W., Harvey, G., Martin, J.H., Parker, P.L., Risebrough, R.W., Robertson, W., Schneider, E. and Gamble, E. (1978): The Mussel Watch. *Environ. Conserv.* 5, 101-125

Hansen, P.D., Bock, R. and Brauer, F. (in press). Investigation of Phagocytosis Concerning the Immunological Defense Mechanism of *Mytilus edulis* using a sub-lethal luminescent bacterial Assay (Photo-bacterium phosphoreum). *J. Comp. Physiol. Biochem.*

Extrapolating Test Results on Bioaccumulation between Organism Groups

Desmond W. Connell

9.1 Abstract

The types of compounds which exhibit least biodegradation and neutral interactions with biota lipid are the most consistent in their bioaccumulation behaviour. Thus it could be expected that the chlorohydrocarbons and polyaromatic hydrocarbons would be the most suitable compounds with which to extrapolate bioaccumulation behaviour between groups of biota. In addition the biota exhibiting least biodegradation and direct partitioning with the external environment are likely to have the most comparable bioaccumulation. Thus bioconcentration behaviour with aquatic organisms can generally be predicted by relationships with the octanol to water partition coefficient. This can be extended to some aquatic infauna and terrestrial invertebrates particularly worms. Patterns of bioaccumulation by terrestrial organisms cannot be predicted from behaviour of aquatic organisms.

9.2 Introduction

When chemicals are discharged to the environment a variety of processes occur which result in distribution to biota which are summarised in Figure 9.1. Firstly it's important to note that a set of abiotic partitioning equilibria are developed which are quantitatively of major importance. These are shown in Figure 9.1 as the water to sediment process, the air to water process and the air to soil process. The water to sediment process can be reasonably well understood as a partition process which can be characterised by the partition coefficient K_d, and in terms of organic carbon, K_{oc} (Karickhoff 1985). The air to water process has been subject to a long period of laboratory investigation and can be characterised by the Henry's Law Constant which is the air to water partition coefficient. The air to soil process is relatively more complex than the others and at the present time is not clearly understood.

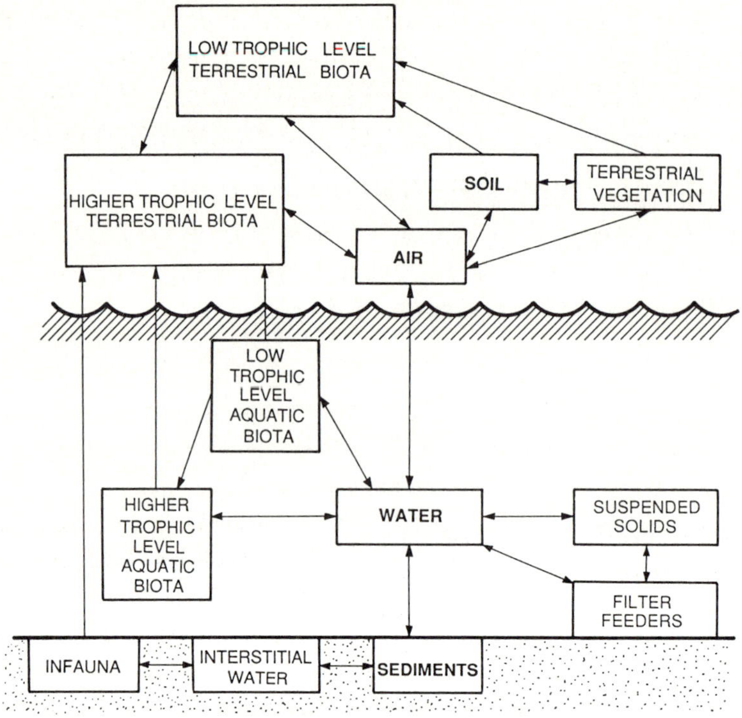

Figure 9.1 Pathways of distribution of a persistent xenobiotic chemical among the biotic and abiotic components of the natural environment.

After distribution by the abiotic processes described previously biota are exposed to compounds in the air, water, interstitial water and soil. Current evidence suggest that some biota bioaccumulate compounds as a result of direct partitioning with the abiotic environment. For example in the aquatic environment the aquatic biota to water and infauna to interstitial water processes can usually be considered to be direct partition processes. In the terrestrial environment some of the low trophic level terrestrial biota to soil and terrestrial vegetation to air as well as terrestrial vegetation to soil processes appear to be direct partition processes. These systems would then be expected to be characterised by a partition coefficient. This would allow the partition approach to be used to predict bioaccumulation.

In actual practice unequal attention has been addressed to these different systems. In fact the aquatic biota to water systems have received the most attention and within this group the fish to water system has received by far the most experimental investigation. Terrestrial systems of bioaccumulation have received comparatively little attention and so the evidence for the operation of partition processes in these systems is not so clear but may be shown by further investigation in the future.

Other chemical transfer processes do not seem to be governed by equilibrium partition processes with the external environment. These occur in the aquatic and terrestrial environments and result mainly where chemicals are transfered to biota in food. The available evidence suggests that partitioning is involved in these processes but they are more complex than with those processes involving direct partition with the external environment.

The objectives of this paper are to evaluate the overall bioaccumulation processes with biota and utilise similarities in these processes as a basis for extending relationships for evaluating bioaccumulation from one group of organisms to another.

9.3 Influence of Type of Compound on Bioaccumulation

A reasonable level of persistence is required for bioaccumulation to occur in biota. Persistence must be such as to allow the concentration of the chemical to accumulate above the level in the external environment. If a compound is subject to biodegradation then the bioaccumulation would be expected to be reduced in accord with the amount of loss of compound which results. Thus for example alkanes exhibit little bioaccumulation despite having all suitable physicochemical properties for this phenomena because they lack persistence in biota. The group with most suitable persistence is the chlorohydrocarbon group and to a lesser extent the group of polyaromatic hydrocarbons.

Bioaccumulation is related to the K_{ow} value when octanol resembles the properties of biota lipid. Generally octanol is a reasonable match for lipid for compounds having log K_{ow} values from 2 to about 6 and these compounds are referred to here as lipophilic compounds. But the different chemical nature of octanol and biota lipid may result in differences in bioaccumulation even with compounds having log K_{ow} between 2 and 6. This could be due to the presence of specific active chemical groups in the test compounds which interact with octanol and lipid in different ways. Molecular dimensions may also influence the bioaccumulation process. In considerations of these factors also, the chlorohydrocarbons and polyaromatic hydrocarbons are relatively neutral compounds and exhibit most predictable behaviour. Connell (1990) has summarised the properties of compounds which influence their bioaccumulation behaviour.

This indicates that relationships for bioaccumulation will be most consistent between organisms and groups of organisms with the chlorohydrocarbons and to a lesser extent the polyaromatic hydrocarbons. Less success can be expected with other groups in making comparisons between different biota. This applies particularly with those which exhibit significant biodegradation.

9.4 Variation in Bioaccumulation with Biological Group

With microorganisms various authors, eg. Kerr and Vass (1973), Sondergren (1968), have known that lipophilic compounds absorb onto the outer surface and then diffuse internally within the cell. Baughman and Paris (1981) have reviewed the available information and concluded that bioconcentration of lipophilic compounds by microorganisms occurs as a result of partitioning between water and the microorganism. With zooplankton investigations by Hardy and Vass (1977) and others have indicated that a related partitioning mechanism occurs with the lipophilic compound being taken up through the carapace. With fish a variety of authors have shown that gills are the site of uptake and partitioning of lipophilic compounds as reviewed by Connell (1988). The evidence indicates that the route of uptake and loss of lipophilic compounds with aquatic organisms is generally through the oxygen uptake route then partitioning with the circulatory fluid occurs resulting in deposition of the lipophilic compound in biota lipid.

The evidence available suggests that bioaccumulation with aquatic organisms is dominated by the bioconcentration process (Connell 1988). The common mechanism of uptake and loss through the oxygen pathway suggests that all aquatic organisms can be treated similarly as regards the water to organism partition process. An outline of this process is contained in Figure 9.2. This indicates that biota fat to water partitioning is the dominant factor in the physical process resulting in the uptake and accumulation of lipophilic compounds from water by aquatic organisms.

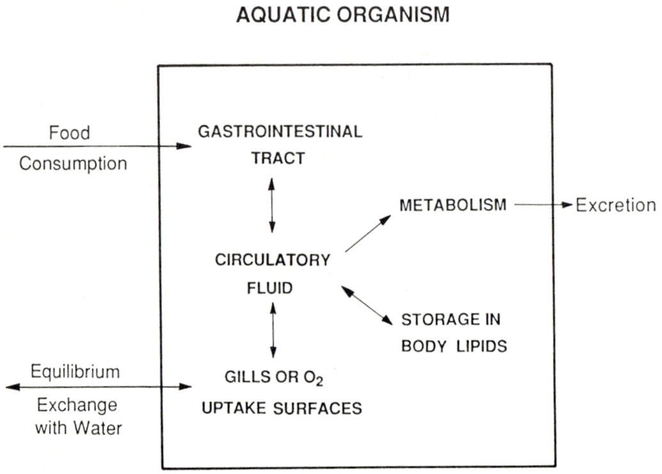

Figure 9.2 A diagrammatic illustration of the patterns of bioaccumulation of a xenobiotic chemical in an aquatic organism.

Apart from the physical process outlined above the persistence of a chemical in an organism is a major factor influencing its capacity for bioaccumulation. The uptake of lipophilic compounds by biota usually results in the induction of mixed function oxidase (MFO) enzyme systems in the exposed biota. The MFO systems stimulate oxidation of lipophilic compounds resulting in the formation of oxidised products which are generally more water soluble than the parent compound and can be more easily excreted and thus removed from the organism. Not all organisms have the same capacity to respond to exposure to lipophilic compounds in this way. The capacity of organisms to respond by MFO induction is important in the bioaccumulation process. Zitco (1980), in reviewing the available information, concluded that metabolic degradation of lipophilic compounds is more rapid in terrestrial biota as compared with aquatic biota. This is reflected in the observation that the main excretion product in aquatic biota is the unmetablised parent compound. This reflects the relatively slow process of removal of the lipophilic compound from biota by the physical partitioning process between the biota and water. However with terrestrial organisms the main excretion products are the oxidised forms of the parent compound which are more rapidly excreted.

Moriarty and Walker (1987) have suggested that these characteristics have developed because aquatic biota have large volumes of water into which lipophilic compounds can be distributed by the physical partitioning process. This process is not available to terrestrial biota which have had to develop oxidation mechanisms to remove relatively large quantities of lipophilic compounds as the oxidised, water soluble products in excretion. These characteristics are illustrated by the data in Figure 9.3 which shows the difference in monooxygenase activity with body size for mammals and fish. This indicates that mammals have considerably higher monoxygenase activity than fish as a general rule. This situation may not apply with all terrestrial biota. For example earthworms, as discussed later, may share characteristics with aquatic biota rather than terrestrial biota. Also Walker et al. (1984) have suggested that some bird species may have low monooxygenase enzyme activity with is more comparable to the level in fish than mammals.

Aquatic organisms can show considerable variation in MFO activity with different species (eg. Connell and Miller 1981). This may relate to particular metabolic characteristics of the species or may relate to factors such as previous exposure patterns to lipophilic compounds. Such differences in MFO activity may lead to different bioaccumulation characteristics for different species.

The previous discussion suggests bioaccumulation relationships established for some aquatic organisms may be generally extrapolated to other organisms. However this must be done with caution since there may be groups of organisms where the relationships don't apply due to different metabolic capacities. The evidence also suggests that relationships for aquatic biota will not be easily applied to terrestrial biota. This is due to the widely different biode-

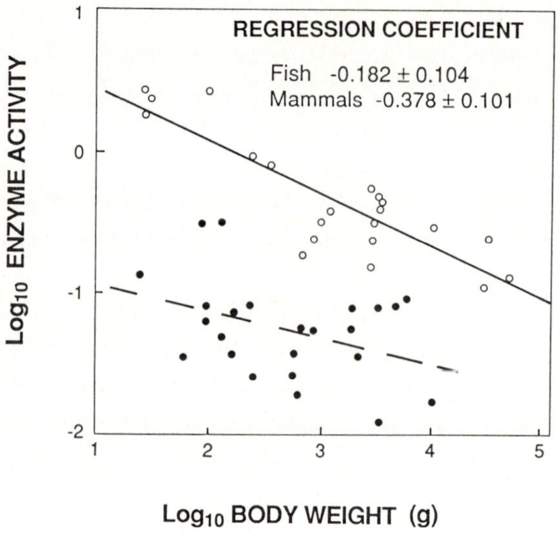

Figure 9.3 Linear regressions for a range of mammalian (O) and fish (●) species enzyme activity and body weight. The activity of the mono-oxygenases associated with liver microsomes is expressed as a proportion of the value for the rat. (From Moriarty, F and Walker, CH, Ecotoxicol. Environ. Saf., 13, 208, 1987. Copyright Academic Press. With permission.)

gradation capacity of the two groups of organisms. Thus generally it would be expected that terrestrial biota would exhibit relatively low capacity to bioaccumulation lipophilic compounds as compared to aquatic biota. In addition the actual bioaccumulation mechanism for terrestrial organisms has not been clearly established but there is little doubt that it will be markedly different from that operating with aquatic organisms. Thus for these two reasons bioaccumulation characteristics of aquatic and terrestrial organisms can be expected to be quite different.

9.5 Bioconcentration by Aquatic Organisms

Bioconcentration is characterised by the bioconcentration factor (K_B) which is the ratio of the concentration of the compound in biota (C_B) to the concentration in water (C_W). An outline of the processes involved in bioconcentration is shown in Figure 9.2. The information generally available indicates that bioconcentration and the relationships derived from this process dominate the bioaccumulation of lipophilic compounds by aquatic organisms. The process is complex but largely governed by the biota lipid to water partition process. As a result it has been found that the most useful characteristic for the prediction of the bioconcentration is the octanol to water partition coefficient (K_{ow}). Despite the simplicity of the octanol to water system as compared with the bioconcentration process the K_{ow} value provides a reasonably good estimate of the bioconcentration capacity of lipophilic compounds within certain limits. The success of the system is largely dependent on the similarity of octanol to biota lipid since water is common to both systems. It is important to note that a variety of other factors, for example the size of the molecule, may be important in the bioconcentration process also.

Compounds which have log K_{ow} values less than 2 usually bioconcentrate more than would be expected from the K_{ow} values. On the other hand, compounds with log K_{ow} values greater than 6 bioconcentrate less than expected. Compounds having log K_{ow} values less than 2 bioconcentrate to a lesser extent due to nonlipoid tissue becoming increasingly important due to the decreasing lipophilicity of these compounds. With compounds having log K_{ow} values greater than 6 the biota lipid and octanol differ in their solubility properties resulting in a reduction of the solubility of these compounds in lipid and a reduced bioaccumulation factor. As mentioned previously in this particular region, properties such as the size of the molecule may also be a factor causing reduced bioconcentration.

If octanol is a perfect surrogate for biota lipid, then

$$K_B = K_{ow}\, y_L \qquad (1)$$

and

$$\log K_B = 1 \log K_{ow} + \log y_L \qquad (2)$$

where y_L is the fraction of lipid present in the biota.

The empirical results obtained from bioconcentration experiments are usually expressed in the following general form:

$$\log K_B = a \log K_{ow} + b \qquad (3)$$

and

$$K_B = 10^b\, K_{ow}{}^a \qquad (4)$$

where a and b are empirical constants.

Based on the outline above constant a is an empirical constant expressing the nonlinearity of the relationship which indicates how well octanol represents the biota lipid, and

$$b = \log y_L$$

This indicates that under perfect conditions y_L is always less than unity which means that constant b should always be negative. Thus organisms with 10% lipid (y_L is 0.1) would give a perfect bioconcentration equation as

$$\log K_B = 1 \log K_{ow} - 1.00$$

Similarity with 1% lipid (y_L is equal to 0.01), then

$$\log K_B = 1 \log K_{ow} - 2.00$$

The observed K_B to K_{ow} relationships with microorganisms are shown in Table 9.1. The relationship reported by Baughman and Paris (1981) may be the most representative of the bioconcentration process since it involved an extensive review of the literature. The constant a values are mostly close to unity suggesting that octanol is a reasonable surrogate for biota lipid. The constant b values are difficult to interpret but the relationship reported by Baughman and Paris would represent 44% lipid content in the dry weight of the microorganism.

Table 9.1 Characteristics of the Relationships between Log K_B and Log K_{ow} for Bioconcentration of Lipophilic Compounds by Microorganisms

Constant a	Constant b	Number of values	r^2	Basis for K_B	Compound	Organisms
0.70	-0.26	8	0.93	Wet wt.	Pesticides	Alga
0.68	+0.16	41	0.81	Wet wt.	Diverse Organic	Alga
1.08	-1.30	8	0.98	Dry. wt.	Condensed ring aromatics	Mixed microbial
0.91	-0.36	14	0.95	Dry wt.	Diverse Organic	Mixed microbial
0.46	+2.36	8	0.83	Wet. wt.	Hydrocarbons and chlorohydrocarbons	Alga
0.36	+2.1	28	0.91	Wet. wt.	Mainly chloro-hydrocarbons	Alga

NOTE: Relationships between log K_B and log K_{OW} takes the form $\log K_B = a \log K_{OW} + b$.

FROM: Connell 1990

There is limited data available for crustacea and this is shown in Table 9.2. The values of constant a are 0.75 and 0.90 with constant b for the relationship reported by Hawker and Connell (1986) representing 5% of lipid wet weight.

Table 9.2 Characteristics of the Relationship Between Log K_B and Log K_{ow} for Bioconcentration of Lipophilic Compounds by Daphnia pulex

Factor	Constant a	Constant b	Number of values	r^2	Basis for K_B	Compound types
log K_{OW}	0.75	-0.44	7	0.85	Wet. wt.	PAH
log K_{OW}	0.90	-1.32	22	0.96	Wet. wt.	Diverse organic

NOTE: Relationship between log K_B and various factors takes the form a log K_B + b.

FROM: Connell 1990.

The large volume of data available on fish is reflected by the equations reported in Table 9.3. Connell and Hawker (1988), Davies and Dobbs (1984), Mackay (1982), Connell and Schüürmann (1988) and Schüürmann and Klein (1988) have collected and collated sets of data and in some cases evaluated the accuracy and application of this material to the bioconcentration relationship. These relationships in most cases were derived from the use of chlorohydrocarbons and polyaromatic hydrocarbons and yielded constant a values of 0.94, 0.98, 1.00, 0.95 and 0.76 with constant b values of -1.0, -1.36, -1.32, -1.06 and -0.35 respectively. These constant b values correspond with lipid contents expressed as a percentage of 10, 4.3, 4.8, 8.8, 47, 23, 14 and 27, Kenaga and Goring (1980) have reported that the lipid content of fish ranges from 1 to about 16% depending on a variety of factors. The relationships reported for molluscs are summarised in Table 9.4. There is limited data available but the application of the alkyl dibenzothiophenes to these current considerations is doubtful because of the presence of active groups in these compounds. The data of Geyer et al. (1982), Hawker and Connell (1986) are most applicable in the current discussion and have constants which are reasonably consistent with octanol providing an acceptable representation

Table 9.3 Characteristics of the Relationships Between Log K_B and Log K_{OW} and for Bioconcentration of Lipophilic Compounds by Fish

Constant a	Constant b	Number of values	r^2	Basis for K_B	Compound types	Year
0.54	+0.12	8	0.95	Wet weight	Organic nonpolar	1974
1.16	-0.75	9	0.98	Wet weight	Various organic	1975
0.63	+0.73	11	0.79	Wet weight	-	1975
0.64	+0.73	11	0.79		-	1978
0.85	-0.70	55	0.95	Wet weight	Various organic	1979
0.94	-1.95	26	0.87	Wet weight	Various organic	1980
0.77	-0.97	36	0.76	Wet weight	Various organic	1980
0.46	+0.63	25	0.63	Wet weight	Various organic	1980
0.83	-1.71	8	0.98	Wet weight	Pesticide	1980
0.98	-0.06	6	0.99	Lipid wt.	-	1980
0.74	-0.77	40	-	-	-	1981
1.00	-1.32	36	0.97	Wet weight	Various organic	1982
1.02	-1.82	9	0.98	Wet weight	Phenols corrected for ionization	1982
1.02	-0.63	11	0.99	Wet weight	Chlorohydrocarbons	1983
0.79	-0.40	122	0.93	-	-	1983
0.94	-0.68	18	0.95	Wet weight	Chlorohydrocarbons	1984
0.60	+1.89	31	0.75	Wet weight	Various organic	1984
0.98	-1.36	20	0.90	Wet weight	Hydrocarbons and chlorohydrocarbons	1984
0.71	-0.92	17	0.98	Wet weight	Aromatic compounds	1984
1.09	-0.87	11	0.9	Wet weight	Chlorohydrocarbons	1985
0.96	-0.56	16	0.98	Wet weight	Chlorohydrocarbons	1985
0.89	+0.61	18	0.95	Lipid wt.	Chlorohydrocarbons	1985
0.96	+0.25	18	0.96	Lipid wt.	Chlorohydrocarbons	1985
0.61	+0.69	11	0.84	-	-	1985
0.94	-1.19	49	0.89	Wet weight	Various organic	1988
0.95	-1.06	30	0.99	Wet weight	Chlorohydrocarbons and PAH	1988
0.94	-1.00	33	0.85	Wet weight	Chlorohydrocarbons and related compounds	1988
0.75	-0.32	32	0.8	Wet weight	Various organic	1988
0.78	-0.35	22	0.95	Wet weight	Chlorohydrocarbons and PAH	1988

NOTE: Relationships between log K_B and log K_{OW} = a log K_{OW} + b.
FROM: Connell 1990.

Table 9.4 Characteristics of the Relationships Between Log K_B and Log K_{ow} for Lipophilic Compounds by Molluscs

Constant a	Constant b	Number of values	r^2	Basis for K_B	Compound Types	Organism	Year
0.86	-0.81	16	0.96	Wet wt.	Various organic	Mussel	1982
0.16	1.52	14	0.71	Wet wt.	Alkyldibenzo thiophenes	Short-necked clam	1984
0.49	1.03	14	0.62	Wet wt.	Alkyldibenzo thiophenes	Oyster	1984
0.31	1.63	14	0.64	Wet wt.	Alkyldibenzo thiophenes	Mussel	1984
0.84	-1.23	34	0.83	Wet wt.	Mainly chloro hydrocarbons	Mollusc	1986

NOTE: Relationships take the form log K_B = a log K_{OW} + b.
FROM: Connell 1990.

of biota lipid. The results suggest that with effectively nonbiodegradeable compounds, principally the chlorohydrocarbons and polyaromatic hydrocarbons, octanol provides a reasonable representation of biota lipid for compounds with K_{ow} values between 2 and 6. The relationships which appear to be most applicable to aquatic biota a summarised in Table 9.5.

The general relationships which can be applied to aquatic organisms take the form indicated in Equations 3 and 4. The results show that constant a may vary from 0.84 to 1.00 depending on the organisms and the particular compounds involved. The value of constant b is usually approximately in accord with the lipid content of the biota being investigated. However it must be kept in mind that particular organisms having different biodegradation capacities may lead to deviations from the general relationship outlined above. Also the presence of functional groups may lead to deviations from the observed relationships which are most applicable to the chlorohydrocarbons and polyaromatic hydrocarbons.

9.6 Bioaccumulation by Aquatic Infauna

Aquatic infauna are fauna which reside within the bottom sediments in aquatic areas. This includes fauna such as aquatic worms and various other organisms. Bioaccumulation with this group can be best understood by a three phase model as outlined in Figure 9.4. This indicates that there are two partition processes involved in bioaccumulation within the sedimentary

Table 9.5 Characteristics of the Most Applicable Relationships Between Log K_B and Log K_{ow} for Various Biota and the Sediment-to-Water System

Biota	Constant a	Constant b	Lipid equivalent to constants b	Actual Lipid (%)	Range of log K_{OW}
Microorganisms	0.91	-0.36	44 (dry wt.)	-	3-7
Daphnids	0.90	-1.32	4 (wet wt.)	-	2-8
Poly- and oligochaetes	0.99	-0.60	25 (dry wt.)	-	4-8
Fish	0.94	-1.00	10 (wet wt.)	1-16 (wet wt.)	3-6
Fish	0.98	-1.36	4.3 (wet wt.)	1-16 (wet wt.)	1.5-6.5
Fish	1.00	-1.32	4.8 (wet wt.)	1-16 (wet wt.)	0.5-6.0
Fish	0.95	-1.06	8.8 (wet wt.)	1-16 (wet wt.)	2-6
Molluscs	0.84	-1.23	5.9 (wet wt.)	1.2-1.8(wet wt.)	3.5-8
Sediment water system	0.99	-0.35	-	-	-
Sediment water system	1.00	-0.32	-	-	-
Sediment water system	1.00	-0.21	-	-	2-6
Sediment water system	0.72	+0.49	-	-	3-5

NOTE; Log K_B (or Koc) = a log K_{OW} + b.
FROM: Connell 1990.

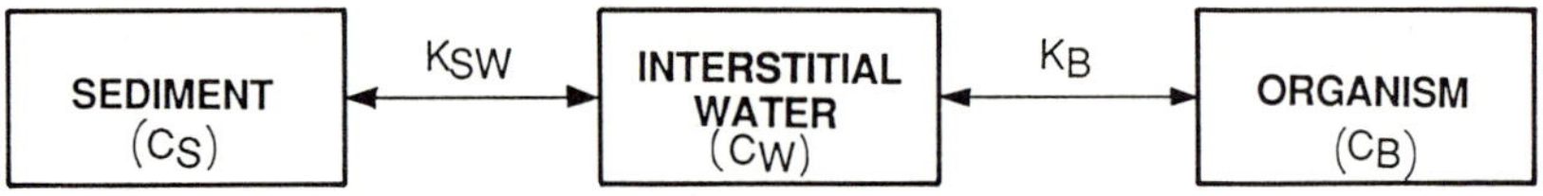

Figure 9.4 Infauna can be considered to be part of a sediment, water and organism system, as illustrated above.

system. Compounds are released from the sediment by the sediment to pore water partitioning system and then there is uptake and accumulation by the pore water to infauna partition process. The sediment to water process is characterised by the equation

$$K_D = C_S/C_W = x\, f_{oc}\, K_{ow}{}^m \qquad (5)$$

where K_D is the sediment to water partition coefficient, x a proportionality constant, m and nonlinearity constant and f_{oc}, the fraction of organic carbon, present in the sediment.

The infauna to water partition process is characterised by the normal bioconcentration relationship where if octanol perfectly represents the biota lipid and equilibrium has been attained, then

$$K_B = C_B/C_{WI} = 10^b\, K_{ow}{}^a = y_L\, K_{ow}{}^a \qquad (6)$$

where C_B is the biota concentration and C_{WI} the concentration in the pore water.

It has been shown by Markwell et al. (1989); and Connell et al. (1988), using Equations 5 and 6, that the bioaccumulation factor, BF, is represented by

$$BF = C_B/C_{sed} = (y_L/x\, f_{oc})\, K_{ow}{}^{a\text{-}m} \qquad (7)$$

The pore water to biota process is bioconcentration and the K_B values should be comparable to the K_B values obtained with other aquatic organisms. However BF is a different characteristic to K_B and is not directly comparable with the bioconcentration process. The nonlinearity constants a and m are often close to unity and the difference between them (a-m) results in a small value or possibly zero being obtained indicating that BF may be independent of K_{ow} or weakly dependent on K_{ow}. The relationship above also suggests that BF will be relatively constant for all lipophilic compounds and primarily dependent on the ratio of the lipid fraction in the biota and the organic carbon content of the sediments.

Markwell et al. (1989) have reported that with aquatic oligochaetes the following bioconcentration equation applies with chlorohydrocarbons

$$\log K_B = 1.11 \log K_{ow} - 1.0 \qquad (8)$$

This is comparable to the bioconcentration relationships which have been established for other aquatic organisms as reported in Tables 9.3 and 9.4. Relationships of this general kind can probably be extended to other aquatic infauna. The evidence provided by Markwell et al. (1989) suggests that the relationship for BF, Equation 7 is probably correct but further evidence would be required to give an unequivocal demonstration of its validity.

9.7 The Soil to Earthworm System

The relationships described above have been relatively successful with aquatic systems but the applications are more limited with terrestrial systems. However the three phase model which has been applied to the sediment to aquatic infauna system has been applied to the soil to earthworm bioaccumulation system by Connell and Markwell (1990). The data used in this work was for a wide variety of chemical groups and also derived using many different experimental systems. The following relationship was found to be applicable to the soilwater to earthworm bioconcentration process

$$\log K_B = 1.0 \log K_{OW} - 0.6$$

The limitations in the data available did not allow a proper evaluation of the BF to K_{OW} relationship (Equation 7). However the data which was available was consistent with the conclusions based on this relationship but were insufficient to demonstrate its validity. Nevertheless this analysis suggests that the three phase system is applicable to earthworms but further confirmatory evidence is needed.

9.8 Bioaccumulation by Terrestrial Biota

A wide variety of transfer processes are possible in terrestrial systems, as outlined in Figure 9.1. Generally terrestrial animals would be expected to exhibit relatively low levels of bioaccumulation as outlined previously. However this applies mainly to mammals and is not appropriate for earthworms, vegetation and birds. Moriarty and Walker (1987) have suggested the following decreasing order of organisms for their capacity to biodegrade lipophilic residues; small vertebrates, large vertebrates, omnivores, herbivores, predators, mammals, birds, fish.

The bioaccumulation process with vegetation can involve two partition processes. These are the atmosphere to above ground foliage and the soil to below ground roots. Generally transfer of compounds to the roots does not lead to further transfer from the roots to the above ground foliage (Connell, 1990). The concentrations which occur in roots are generally similar or lower than that which occurs in the adjacent soil. A limited amount of data is available on the atmosphere to foliage bioaccumulation process. Gaggi et al. (1985), Bacci et al. (1990) and Riederer (1990) have estimated bioaccumulation factors for several compounds with the atmosphere to foliage partitioning process and shown the value of physicochemical properties in predicting these.

Often birds are found to have the highest concentrations of residual chemicals which are observed in terrestrial systems. Walker et al. (1984) have reported that seabirds are among the highest due to the consumption of aquatic organisms, particularly fish, which are relatively high in chemical residues. The mechanism of this bioaccumulation process is not well understood and extension of relationships for bioaccumulation with other organisms is not possible at present.

9.9 Conclusions

The available evidence indicates that bioaccumulation is relatively well understood and relationships for its estimation have been best established with fish. It is likely that the relationships established with fish can be extended to other aquatic organisms. However there needs to be caution in applying these relationships since they are most applicable with the chlorohydrocarbons and deviations may occur due to chemical factors with other substances. Also these relationships depend on a lack of biodegradation within the organism during the bioaccumulation process. The least biodegradable compounds are the chlorohydrocarbons but biodegradation may occur to different extents with other organisms which may have differing capacities to carry out for this process. As a result different organisms may possibly exhibit different capacities to bioaccumulate various compounds.

The relationships which have been developed for aquatic organisms can be extended to the earthworm bioaccumulation process in soil. This extension may be applicable to other terrestrial invertebrates in soil as well but insufficient evidence is available to confirm this at present. The relationships to evaluate bioaccumulation cannot be extended to mammals due to their relatively biodegradation high capacity. In addition the bioaccumulation process is quite different from that operating with aquatic organisms. In general this means that these organisms have relatively low capacity to bioaccumulate lipophilic compounds. Its likely that the vegetation to atmosphere system can be understood and predicted by the partition approach.

9.10 References

Bacci, E., Calamari, D., Gaggi, C. and M. Vighi (1990). Bioconcentration of Organic Chemical Vapours in Plant Leaves: Experimental Measurements and Correlation. *Environ. Sci. Technol.* 24, 885.

Baughman, G.L, D.F. and Paris (1981). Microbial Bioconcentration of Organic Pollutants from Aquatic Systems - A critical review, *CRC. Cri. Rev. Microbiol*, Jan, 205.

Connell, D.W. (1988). Bioaccumulation Behaviour of Persistent Organic Chemicals with Aquatic Organisms, *Rev. Environ. Contam. Toxicol*, 101, 117.

Connell, D.W. (1990). *Bioaccumulation of Xenobiotic Compounds*, CRC Press, Boca Raton.

Connell, D.W., Bowman, M. and D.W. Hawker (1988). Bioconcentration of Chlorinated Hydrocarbons from Sediment by Oligochaetes. *Ecotoxic. Environ. Safety*, 16, 293.

Connell, D.W. and D.W. Hawker (1988). Use of Polynomial Expressions to Describe the Bioconcentration of Hydrophobic Chemicals by Fish. *Ectoxicol. Environ. Safety*. 16, 242.

Connell, D.W. and M.R.D. Markwell (1990). Bioaccumulation in the Soil to Earthworm System, *Chemosphere*, 20, 99.

Connell, D.W. and G.J. Miller (1981). Petroleum Hydrocarbons in Aquatic Ecosystems - Behaviour and Effects of Sublethal Concentrations. *CRC Rev. Environ. Control*. 11, 37.

Connell, D.W. and G. Schüürmann (1988). Evaluation of Various Molecular Parameters as Predictors of Bioconcentration in Fish, *Ecotoxicol. Environ. Safety*, 15, 324.

Davies, R.P. and A.J. Dobbs (1984). The Prediction of Bioconcentration in Fish, *Water Res.*, 18, 1253.

Gaggi, E.E.C., Bacci, E., Calamari, B. and R. Finelli (1985). Chlorinated Hydrocarbons in Plant Foliage: An Indication of the Tropospheric Contamination Level, *Chemosphere*, 14, 1673.

Geyer, H., Seehan, B., Kotzias, B., Freitag, B., and F. Korte (1982). Prediction of Ecotoxicological Behaviour of Chemicals; Relationship Between Physicochemical Properties and Bioaccumulation of Organic Compounds in the Mussel, *Chemosphere*, 11, 1121.

Hawker, D.W. and D.W. Connell (1986). Bioconcentration of Lipophilic Compounds by Some Aquatic Organisms, *Ecotoxicol. Environ. Safety*, 11, 184.

Karickhoff, S.W. (1985). Sorption Phenomena, in *Environmental Exposure from Chemicals*, Vol. 1, Neely, WB, Blau, GE, Eds. CRC Press, Boca Raton, p. 49.

Kenaga, E.E. and C.A. Goring (1980). Relationship Between Water Solubility, Soil Sorption, Octanol Water Partitioning and Bioconcentration of Chemicals in Biota, in *Aquatic Toxicology*, Eaton, JG, Parrish, PR and Hendricks, AC, Eds., Vol. 707, American Society for Testing and Materials, Philadelphia, p. 78.

Kerr, S.R. and W.P. Vass (1973). Pesticide Residues in Aquatic Invertebrates, in Edwards, CA, Ed., *Environmental Pollution by Pesticides*, Plenum Press, London, p. 134.

Mackay, D. (1982). Correlation of Bioconcentration Factors, *Environ. Sci. Technol.*, 16, 274.

Markwell, R.D., Connell, D.W. and A.J. Gabric (1989). Bioaccumulation of Lipophilic Compounds from Sediments by Oligochaetes. *Water Res.*, 23, 1443.

Moriarty, F. and C.H. Walker (1987). Bioaccumulation in Food Chains - A Rational Approach, *Ecotoxicol. Environ. Safety*, 13, 208.

Riederer, M. (1990). Estimating Partitioning and Transport of Organic Chemicals in the Foliage/Atmosphere System: Discussion of a Fugacity Based Model, *Environ. Sci. Technol.*, 24, 829.

Schüürmann, G. and W. Klein (1988). Advances in Bioconcentration Prediction, *Chemosphere*, 17, 1551.

Sondergren, A. (1968). Uptake and Accumulation of DDT by Chlorella sp., *Oikos* 19, 126.

Walker, C.H., Knight, G.C., Chipman, J.K., and M.J.J. Ronis (1984). Hepatic Microsomal Monooxygenases of Seabirds, *Mar. Environ. Res.*, 14, 416.

Zitco, V. (1980). Metabolism and Distribution by Aquatic Animals, *In Handbook of Environmental Chemistry*, Hutzinger, O, Ed., Springer Verlag, Berlin, p. 221.

Extrapolating the Laboratory Results to Environmental Conditions

Dick T.H.M. Sijm

10.1 Abstract

Bioaccumulation kinetics are usually considered to follow first order one compartment kinetics. This has been determined in many laboratories with different types of compounds. This means that if the kinetic parameters are known, the concentrations of contaminants in fish in the field are only dependent on ambient water concentration, concentration in the food and exposure time. Here, it is discussed if first order one compartment kinetics may be applied to field conditions.

10.2 Introduction

Bioconcentration of organic chemicals by fish has often been modelled (Branson et al. 1975, Neely 1979, Spacie and Hamelink 1982, Banerjee et al. 1984). Usually first order one compartment kinetics have been found to adequately describe bioconcentration. Fish is regarded as one homogeneous compartment in which chemicals may enter from the water with an uptake rate constant (k_1, mL/g/d) or from food with an uptake rate constant (E*f, g_{food}/g_{fish}/d) and leave the fish with an elimination rate constant (k_2, 1/d):

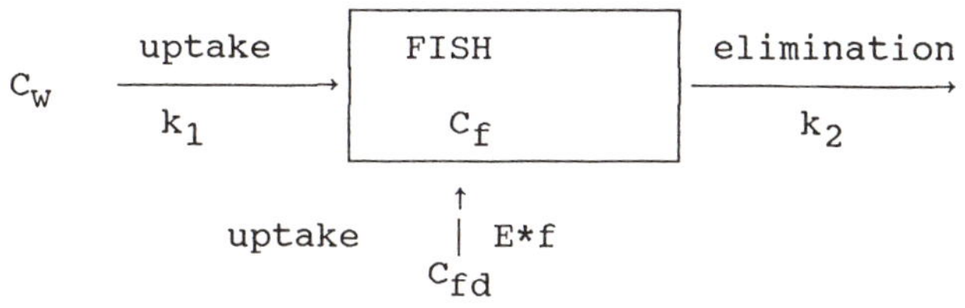

So,

$$dC_f/dt = k_1*C_w + E*f*C_{fd} - k_2*C_f \quad (1)$$

If the concentration in water (C_w) and food (C_{fd}) are constant, the concentration in fish (C_f) can be determined following

$$C_f = [(k_1*C_w + E*f*C_{fd}) / k_2] * [1 - \exp(-k_2*t)] \quad (2)$$

where t is time (d). When uptake of the organic chemicals occurs predominantly via the water phase the bioconcentration constant (K_c, mL/g) can be determined at steady state conditions, i.e. $dC_f/dt = 0$, and

$$K_c = C_f/C_w = k_1/k_2 \quad (3)$$

When uptake occurs predominantly via the food chain the biomagnification factor (K_m) can be determined at steady state conditions ($dC_f/dt = 0$), and

$$K_m = C_f/C_{fd} = E*f/k_2 \quad (4)$$

In long term studies with growing fish growth is an additional factor which causes a decrease of the concentration in the fish, i.e. by growth dilution which can be regarded as a first order dilution rate constant (τ). In equations (1-4) the elimination rate constant can then be expanded to a "real" elimination rate constant (k_2) plus a growth dilution rate constant (τ).

In many studies one compartment first order kinetics have thus been observed for the bioaccumulation of organic chemicals in fish. However, the use of accelerated bioconcentration tests may lead to erroneous conclusions. Biphasic elimination will not be observed in short-term experiments in the laboratory (Stehly and Hayton 1989). Such two-compartment first order kinetics have been described by Spacie and Hamelink (1982) and was for instance experimentally observed in a long-term elimination study of PCBs in guppy in the laboratory by Schrap and Opperhuizen (1988).

10.3 Comparing Laboratory and Field Conditions

All first order kinetic processes have been studied in the laboratory with several types of organic chemicals and different types of fish. To verify if the first order one compartment model is applicable to field situations the laboratory conditions will be critically compared to those from field situations. As many conditions as possible will be reviewed in order to provide an insight to what extent laboratory results may be extrapolated to field conditions. The most important conditions are visualized in Figure 10.1.

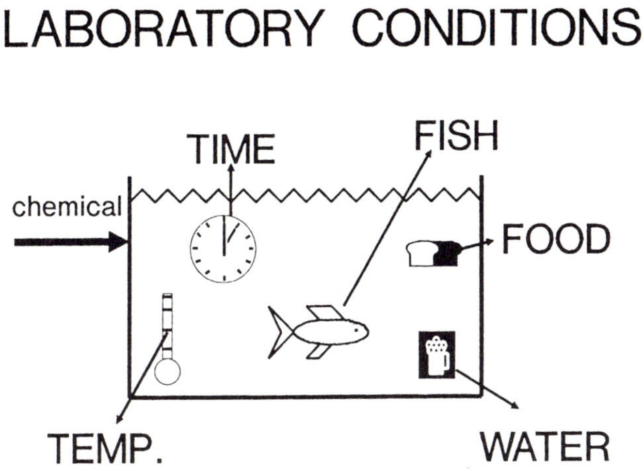

Figure 10.1 Schematic representation of conditions in the laboratory which may have an influence of extrapolating experimental laboratory data to field data.

10.3.1 Fish

In the laboratory often fish are used which have several practical advantages. i) Fish have to be easily bred to get cheap and succesful reproduction. ii) They must not be readily subjected to diseases of any kind in order to prevent mortality in control groups. iii) Often fish are used which have high commercial value, e.g. rainbow trout. iv) Small fish are used which has the advantage that relatively small aquaria can be used, small volumes of water are needed and low amounts of (expensive) chemicals. v) Often tropical fishes are used to follow some of the aforementioned conditions. They have to be maintained at high temperatures (25 °C), the advantage being that working with these fish can be done under milder conditions than the often lower temperatures at field conditions. vi) Often adult fish are used to prevent growth dilution and to simplify kinetics.

Some of these conditions will be discussed in more detail which allows one to get more insight in relation to a possible extrapolation to field conditions.

Species. Bioconcentration factors of 1,2,4-trichlorobenzene have been determined for a number of fishes, which differed between 124 (mL/g) for rainbow trout (1.8% lipid content) and 3,200 (mL/g) for rainbow trout (8.8% lipid content). It was assumed that this hydrophobic chemical is stored in the lipid fraction of the fish and the bioconcentration factors were corrected for lipid content. Bioconcentration factors (on a lipid weight basis) thus obtained, varied less but still between 6,890 and 36,364 (mL/g_{fat}) (Geyer et al. 1985). Davies and Dobbs (1984) also compiled several literature data from different species and found for instance a 14-fold difference for the bioconcentration factors for toxaphene in fathead minnow and brook trout, respectively. They however, did not correct for lipid content of the fish. Simulation studies by Jensen et al. (1982). indicate that differences in PCB concentrations among species and in the same species among different environments result from differences in metabolic parameters, exposure, size, and rate of growth.

These studies suggest that one cannot simply use the bioconcentration factor obtained in the laboratory for one species to predict the bioconcentration factor in the field for the same or for other species.

Age (Growth). In laboratory studies it is often found that equilibrium is reached between concentrations in fish and water, respectively, or in fish and food, respectively. Bruggeman (1983) reported bioconcentration and biomagnification factors and kinetic parameters such as uptake efficiencies and elimination rate constants for PCBs in adult guppy. However, in a recent study it was observed that in young growing guppy uptake efficiencies for PCBs were dependent on age (Sijm and Opperhuizen, unpublished results). Juvenile guppy are not capable to digest food at high extent which consequently leads to a low uptake efficiency of the contaminants from the food. Lieb et al. (1974) showed that growing rainbow trout attained steady state concentrations after feeding them PCB contaminated food. In field studies it appeared that concentrations of PCBs increase with increasing age (Wszolek et al. 1979, Connell 1987) which from a thermodynamic basis is not possible (Connolly and Pedersen 1988). The theoretical explanation for these field observations is that the higher fugacity of chemicals in fish is due to an even higher fugacity of the chemicals in the gut due to food volume reduction compared to food not yet digested (Connolly and Pedersen 1988).

It may be concluded from these studies that data obtained from the laboratory using adult fish cannot always be used for juvenile fish. Also the rate of which fish are growing is very important to establish a steady state concentration of organic chemicals in fish while exposed. This means that fish will reach higher concentrations of contaminants if no growth occurs. To distinguish growth dilution from "real" elimination of the compounds is very difficult under laboratory conditions. In particular for compounds with very low elimination rate constants, such as for the extremely hydrophobic compounds it is hard to distinguish growth dilution and "real" elimination. Elimination rate constants for these extremely hydrophobic compounds are not easy to determine in the laboratory and it will thus be difficult to extrapolate to field conditions.

10.3.2 Food

In the laboratory fish have to be fed in order to keep them alive and healthy. Specific diets have been manufactured to feed fish. However, fish in the natural environment have a different feeding behaviour and sometimes a greater variety of food. Little is known about the influence by the type of food on bioconcentration. Recent results show that the uptake efficiency of hydrophobic compounds depends highly on the type of food, being either very lipid or protein like (Sijm and Opperhuizen, unpublished results). In bluegill sunfish uptake of benzo(a)-pyrene was twice as fast when fish were fed compared to unfed fish, elimination, however, was even more pronounced affected by feeding. Fed fish eliminated benzo(a)pyrene 10 times as fast as unfed fish (Jimenez et al. 1987). This clearly shows that food can have a dramatic influence on bioconcentration kinetics.

10.3.3 Water

The water to which fish are exposed in laboratory experiments may not always reflect natural water. An obvious reason for this is that natural waters already contain too much of contaminants that fish would be exposed to a mixture of compounds even in control groups. However, when the bioconcentration in zebrafish of lindane, 3,4-dichloroaniline, 4-nitrophenol and phenol was studied in fish which were exposed to the chemicals in tap water or river water, no differences could be observed (Ensenbach and Nagel 1990). These laboratory results may thus be extrapolated to field situations.

Due to differences in temperature and other influences the oxygen concentration in environmental waters may differ. Opperhuizen and Schrap (1987), however, showed that for two hydrophobic compounds bioconcentration kinetics were not influenced at oxygen concentrations between 2.5 and 8.0 ppm.

In environmental waters the pH may also differ at different locations. In particular for phenolic and carboxylic acid compounds the degree of dissociation may differ at different pH. Consequently, the uptake of these compounds may be dependent on the pH of the water. Saarikoski et al. (1986) showed these different uptake rates at different pH but could not fully account for the degree of dissociation to explain the differences. Extrapolation of bioconcentration kinetics to field situations will thus be difficult when one has to account for varying pH.

10.3.4 Temperature

Temperatures in laboratory experiments usually are performed at room temperatures, i.e. in the range of 20-25°C. These temperatures are usually higher than those in waters in the environment, dependent on geographic location such as latitude. Since fish are poikilothermic animals a number of physiological processes in the fish are different at lower temperatures which may influence bioconcentration kinetics. Therefore somc studies investigated the role of temperature on bioconcentration kinetics. The uptake and elimination rate constants of 4-amino-antipyrine and ethanol in goldfish increased with increasing temperatures between 10 and 35°C (Kaka and Hayton 1978). DDT and methylmercuric chloride were both taken up in higher extent at higher temperatures in rainbow trout between 5 and 15°C (Reinert et al. 1974). However, in both studies steady state conditions were not reached. Therefore, it is not clear if temperature dependent differences in bioconcentration factors were observed. Opperhuizen et al. (1988) observed only small differences in the bioconcentration factors of polychlorinated benzenes in guppy between 13 and 33°C. Jimenez et al. (1987) observed a 5.8-times higher uptake rate constant and a 3.6-times higher elimination rate constant of benzo(a)pyrene in bluegill sunfish at 23°C compared to 13°C. Benzo(a)pyrene thus accumulates to a higher extent at higher temperature. Due to a lack of data, however, make it impossible to draw general conclusions on the extrapolation from laboratory results to field conditions with regard to temperature.

10.3.5 Time

To obtain kinetic data in the laboratory it is required that one takes enough time to follow the change in concentration in a fish. In particular for extremely hydrophobic compounds, it is sometimes impossible to expand experimental time-scales in order to do so (Hawker and Connell 1988). One need to reach steady state situations to reliably determine the bioconcentration or biomagnification factors. For compounds which have a elimination rate constant of 0.001 d^{-1}, the time to reach 90% of steady state in non growing fish is 3 year. This period is too long to measure this rate constant, in particular since small fish do not live such a time. In addition, over such a period of time growth will be causing dilution of the chemical. When the growth rate constant is approximately 0.002 d^{-1}, 90% of steady state will be reached after 1 year. However, small variations in this growth rate may have great influences on the estimation of the "real" elimination rate constant of this particular compound. Neither in field situations nor in laboratory experiments, steady state will be reached for compounds with extreme low elimination rate constants and it will be very difficult to extrapolate estimated data from laboratory experiments to field situations. Except in growing fish where the growth dilution constant is higher than the elimination rate constant, steady state may be reached. Then the concentration of the chemicals in fish at field conditions can be estimated.

10.3.6. Single/multiple Compounds

Usually the bioconcentration kinetics of single compounds are determined in laboratory experiments. In the field, however, fish are exposed to mixtures of compounds. Therefore there is a need to verify whether the presence of other chemicals may influence the kinetic parameters which have been determined from single compounds studies. Opperhuizen and Jongeneel (1986) have studied the bioconcentration of a PCB mixture, Aroclor 1254, and compared individual components from that mixture to the kinetic parameters of PCB congeners which had been individually examined. They concluded that with regard to the bioconcentration kinetics, components in a mixture do not behave differently from the individual congeners, i.e. the components were all taken up with comparable uptake rate constants and were eliminated from the fish with different elimination rate constants. The composition of the mixture of PCBs after uptake will differ in time from the original Aroclor mixture to which the fish have been exposed and the higher chlorinated congeners will persist the longest time in the fish. This latter phenomenon has also been found in the field study by Wszolek et al. (1979). The laboratory results thus seems to agree with field data.

The presence of other chemicals in the environment may lead to the following processes. i) The solubilities of organic chemicals may be enhanced in the presence of organic solvents (Rao et al. 1990) or lowered in the presence of other chemicals (Opperhuizen and Jongeneel 1986) which may influence the bioaccumulation of these chemicals. ii) The amount of "freely dissolved" chemicals may be decreased due to coagulation or sorption to other chemicals. This also may influence bioaccumulation (Black and McCarthy 1988, Schrap and Opperhuizen 1990).

10.4 Discussion

It has been shown that a number of laboratory conditions at which bioconcentration kinetics are measured have had little influence on the extrapolation of the kinetic data to field situations. Several studies even show good correlations between laboratory and field bioconcentration factors for chemicals with short half-lives (relatively high elimination rate constants). For chemicals with long half-lives or chemicals which were biotransformed by fish no correlations were found (Davies and Dobbs 1984, Oliver and Niimi 1983, Oliver and Niimi 1985).

For a number of conditions, however, an extrapolation from laboratory to field conditions cannot be simply made. For instance fish living in lakes which are frozen in winter usually do not feed during that period. First order kinetics, however, predicts they are fed continuously the whole year through. Also, a pulse exposure of the contaminant or multiple pulses of the contaminants are not included in first order kinetics. How to cope with these kinds of problems are issues of future research.

For a number of studies it has been recognized that bioconcentration is a first order kinetic process. It must however, be beared in mind, that most of these studies have been performed with chlorinated aromatic compounds of relatively low molecular weight (Davies and Dobbs 1984). To generalize the theory on bioconcentration kinetics the range of chemical types should be expanded.

10.5 Conclusions

First order one compartment kinetics which describe the bioaccumulation process as derived in laboratory experiments can be applied to field conditions for most chlorinated organic chemicals with low molecular weight and which are not subjected to biotransformation. It is more difficult to verify if these kinetics can also be applied for other compounds. For highly hydrophobic compounds it is very difficult to obtain kinetic data both in the field and in the laboratory since the time required for obtaining reliable data is too long. For other chemicals little information is available regarding bioaccumulation kinetics.

A number of conditions under which experiments have been performed in the laboratory mimic field conditions. These are water quality, oxygen concentration and kinetics of mixtures vs. single compounds. However, some conditions cannot be simply regarded as having no influence on the data obtained in the laboratory. Bioconcentration kinetics obtained for one fish cannot be extrapolated to other fish or to the same fish under different environmental conditions. Also data obtained from adult fish cannot simply be applied to juvenile fish. The influence of temperature and the type of food is not well investigated. The influence of pH to the degree of dissociation, in particular for phenolic and carboxylic acid compounds, associated to the bioconcentration kinetics is not well understood. The influence of mixtures of compounds on the bioavailability and solubility is also not well understood. Finally, the experimental time in the laboratory may be too short to distinguish between one compartment and two compartment first order kinetics.

10.6 References

Banerjee, S., R.H. Sugatt and D.P. O'Grady (1984). *Environ. Sci. Technol.*, 18, 79-81.

Black, M.C. and J.F. McCarthy (1988). *Environ. Toxicol. Chem.*, 7, 593-600.

Branson, D.R., G.E. Blau, H.C. Alexander and W.B. Neely (1975). *Trans. Amer. Fish. Soc.*, 104, 785-792.

Bruggeman, W.A. (1983). Ph.D. Thesis, University of Amsterdam.

Connell, D.W. (1987). *Chemosphere*, 16, 1469-1474.

Connolly, J.P. and C.J. Pedersen (1988). *Environ. Sci. Technol., 22*, 99-103.

Davies, R.P. and A.J. Dobbs (1984). *Water Res.*, 18, 1253-1262.

Ensenbach, U. and R. Nagel (1991). *Comp. Biochem. Physiol.*, 100 C, 49-54.

Geyer, H., I. Scheunert and F. Korte (1985). *Chemosphere*, 14, 545-555.

Hawker, D.W. and D.W. Connell (1988). *Wat. Res.*, 22, 701-707.

Jensen, A.L., S.A. Spigarelli and M.M. Thommes (1982). *Can. J. Fish. Aquat. Sci.*, 39, 700-709.

Jimenez, B.D., C.P. Cirmo and J.F. McCarthy (1987). *Aquatic Toxicol.*, 10, 41-57.

Kaka, J.S. and W.L. Hayton (1978). *J. Pharm. Sci.*, 67, 1558-1563.

Lieb, A.J., D.D. Bills and R.O. Sinnhuber (1974). *J. Agric. Food Chem.*, 22, 638-642.

Neely, W.B. (1979). *Environ. Sci. Technol.*, 13, 1506-1510.

Oliver, B.G. and A.J. Niimi (1983). *Environ. Sci. Technol.*, 17, 287-291.

Oliver, B.G. and A.J. Niimi (1985). *Environ. Sci. Technol.*, 19, 842-849.

Opperhuizen, A. and R.P. Jongeneel (1986). In: Organic Micropollutants in the Aquatic Environment, Eds.: A. Bjorseth and G. Angeletti, D. Reidel Publishing Company, Dordrecht, The Netherlands, pp. 251-260.

Opperhuizen, A. and S. M. Schrap (1987). *Environ. Toxicol. Chem.*, 6, 335-342.

Opperhuizen, A., P. Serné and J.M.D. Van der Steen (1988). *Environ. Sci. Technol.*, 22, 286-292.

Rao, P.S.C., L.S. Lee and R. Pinal (1990). *Environ. Sci. Technol.* , 24, 647-654.

Reinert, R.E., L.J. Stone and W.A. Willford (1974). *J. Fish. Res. Board Can.*, 31, 1649-1652.

Saarikoski, J., R. Lindström, M. Tyynelä and M. Viluksela (1986). *Ecotox. Environ. Safety*, 11, 158-173.

Schrap, S.M. and A. Opperhuizen (1988). *Bull. Environ. Toxicol. Chem.*, 40, 381-388.

Schrap, S.M. and A. Opperhuizen (1990). *Environ. Toxicol. Chem.*, 9, 715-724.

Stehly, G.R. and W.L. Hayton (1989). In: Aquatic Toxicology and Environmental Fate: Eleventh Volume, ASTM STP 1007, Eds.: G.W. Suter II and M.A. Lewis; American Chemical Society for testing and Materials, Philadelphia, pp. 573-584.

Spacie, A. and J.L. Hamelink (1982). *Environ. Toxicol. Chem.*, 1, 309-320.

Wszolek, P.C., D.J. Lisk, T. Wachs and W.D. Youngs (1979). *Environ. Sci. Technol.*, 13, 1269-1271.

Bioaccumulation: Does it Reflect Toxicity?

James M. McKim and Patricia K. Schmieder

11.1 Abstract

Toxicity/bioconcentration-based residue estimates (TBRE), a combination of toxicity QSARs and bioconcentration QSARs, have provided good first approximation estimates of toxic internal residues for those chemicals which have a non-specific, passive, toxic mode of action such as non-polar narcosis. Data on measured and estimated residues reviewed here for several species of fish support the TBRE approach for non-reactive chemicals. Measured and estimated toxic residues for several other toxic modes of action (polar narcosis, uncouplers, skin irritants, acetylcholinesterase inhibitors, respiratory blockers, and central nervous system seizure agents) also support the TBRE approach. The toxic residues observed for these other modes of action measure lower than those related to non-polar narcosis and there seems to be a clear delineation in toxic residues between the established modes of action. Those chemicals which exhibit a specific mode of action demonstrate a strong trend towards similiar potency regardless of their log K_{ow}. Since each mode of action identified to date has shown a characteristic toxic residue, it seems important, to ensure sensitivity and accuracy of residue estimates, that TBRE be developed (as are aquatic toxicity QSARs) for chemicals which represent specific toxic modes of action. At this time, TBRE which have been developed for chemicals within specific toxic modes of action may provide aquatic toxicologists with the best first approximation of a toxic residue.

11.2 Introduction

In the past decade the field of ecotoxicology has moved with increasing enthusiasm toward the development of sophisticated approaches to predictive toxicology. Demands are great on governments worldwide to regulate some 50,000 chemicals already in production as well as hundreds of new chemicals going into production every year that could impact the environment. Regulatory toxicology efforts that attempt to protect ecosystems from the potential adverse impacts of this vast chemical inventory must of necessity incorporate a strong predictive capability. At this time, the most comprehensive approach to toxicity predictions employs quantitative structure activity relationships (QSAR).

QSAR methods probe the interrelationships between the molecular structure of chemicals, as described by specific physical-chemical properties, and biological activity. The use of QSAR in aquatic toxicology has been developing rapidly over the last decade following the pioneering efforts of Veith et al. (1979), Konemann (1981), and Veith et al. (1983). These early studies revealed the relationship between octanol-water partition coefficients (K_{ow}), and bioconcentration and/or acute toxicity in fish. A review on QSAR by Hermens (1989) describes a rapid expansion of this early predictive approach to other chemicals and species. These developments have made QSAR an indispensable tool for predicting and regulating the effects of environmental chemicals.

Recently, Friant and Henry (1985) proposed that measured internal concentrations of chemicals that coincide with specific toxic endpoints in aquatic species would provide a better regulatory tool than aqueous chemical concentrations because bioavailability problems could be eliminated. (McCarty 1986, 1987a, b) has extended the use of aqueous toxicity QSARs by his demonstration that combinations of bioconcentration and toxicity QSARs can provide an estimate, within an order of magnitude, of internal chemical concentrations in fish at acute and chronic endpoints for narcotic and some non-narcotic industrial chemicals. Donkin et al. (1989) extended this work to relationships between bioconcentration, hydrophobicity, and sublethal toxic endpoints in a marine mussel that results in toxic internal tissue concentrations causing a specific biological response. Further support for use of tissue chemical residue in predictive toxicology was offered by Van Hoogen and Opperhuizen (1988), who used a first-order bioconcentration model to predict exposure concentrations that cause lethal residues in the guppy at specific exposure times. They further measured lethal tissue concentrations to validate the model predictions. These studies all support the contention that the internal concentration of a chemical is a good surrogate for the chemical concentration at the site of toxic action. Furthermore, their work indicates that chemicals with a non-polar narcosis toxic mode of action show equal potency on the basis of internal tissue concentrations that occur at specific toxic endpoints.

The purpose of this paper is to: (1) present a synopsis of new ideas and assumptions on the relationships between toxicity and bioconcentration that propose to combine QSARs for bioconcentration and toxicity (acute and chronic) into a predictive tool for toxic body burdens in aquatic species, (2) provide some data sets on toxic body burdens in fish that may help to further evaluate the approach, (3) discuss some of the attributes and problems with the present use of a QSAR-predicted toxic body burden, and (4) recommend several research areas critical to further development of QSAR-based toxic body burden determinations.

11.3 QSAR Approach in Aquatic Toxicology

11.3.1 Bioconcentration QSAR

The first use of QSAR in aquatic toxicology involved the relationship between the octanol-water partition coefficient (log K_{ow}) and the bioconcentration of waterborne chemicals (Veith et al. 1979). At the same time, this work provided a rapid liquid chromatography method for accurately determining the log K_{ow} of organic chemicals. This physical-chemical information on log K_{ow} for each chemical coupled with empirically derived steady-state tissue concentrations provided a relationship (Veith et al. 1979) that allowed the prediction of total body residues (Figure 11.1). Mackay (1982) developed a mechanistic model to describe the log k_{ow}/bioconcentration relationship, while Veith and Kosian (1983) added more species and chemicals to better define their original empirical equation.

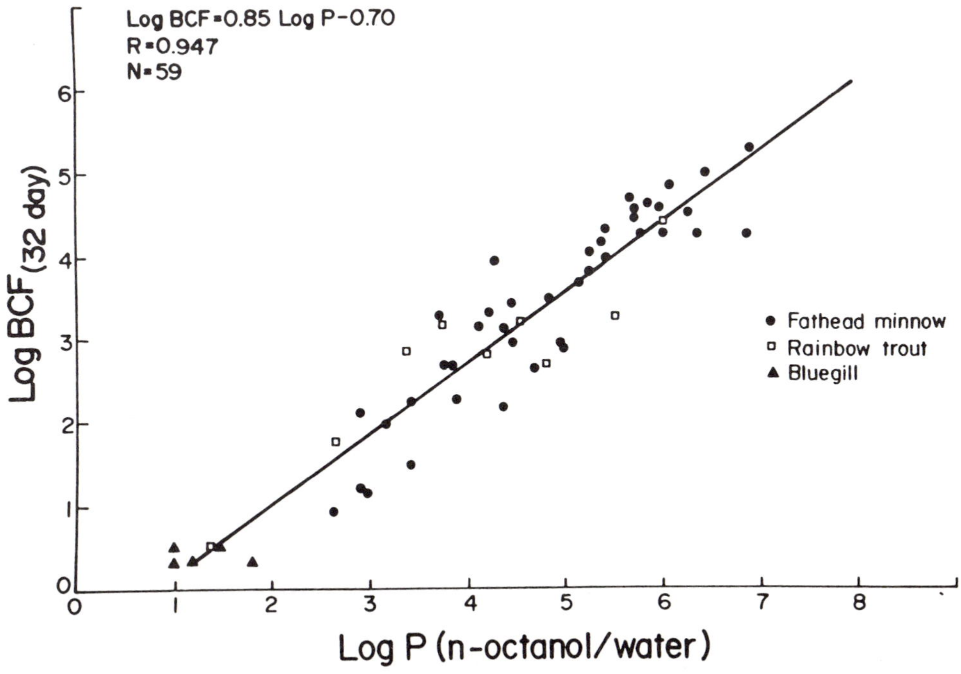

Figure 11.1 Correlation of log BCF with log P for organic chemicals.

11.3.2 Toxicity QSAR

Koenemann (1981), Veith et al. (1983), and Veith et al. (1985) began experimentation with the use of QSAR in estimating the acute toxicity of industrial chemicals to guppies and fathead minnows, respectively. This work was a continuation of some of the basic findings of Ferguson (1939) who showed that non-specific acute toxicity (narcosis) was directly proportional to log K_{ow}. The work on fathead minnows involved 65 industrial chemicals representing 4 major chemical classes (ketones, ethers, alkyl halides, and substituted benzenes) and strongly established a narcosis acute toxicity QSAR for fish. This non-polar narcosis QSAR is a predictor of baseline acute fish toxicity (Veith et al. 1983; Koenemann, 1981) and is representative of a major portion of the industrial chemicals impacting the aquatic environment today (Figure 11.2). This early fish work was followed by initiation of QSAR development for several other animal species and several new chemical classes (Calamari et al. 1983; Rogerson et al. 1983; Hermens et al. 1984a; McLeese et al. 1979; Ribo and Kaiser 1983; Schultz et al. 1986; VanGestel and Ma 1988) (Table 11.1). The results are quite similar for specific classes of non-polar narcotic chemicals and for several different animal species (Hermens 1989).

The narcosis QSAR is currently considered a predictor of baseline toxicity and lies within the space delineated by the relationship between water solubility and log K_{ow}. This "Aquatic Toxicity Space" occupies the area below the water solubility line in Figure 11.2. Although the majority of industrial chemicals can be treated as narcotics, the toxicity of compounds associated with other more specific modes of action (e.g., acetylcholinesterase (AChE) inhibition, uncoupling of oxidative phosphorylation, etc.) would be underestimated with the narcosis QSAR; i.e., the toxicity QSARs for these more specific acting chemicals lie below the non-polar narcosis QSAR in "Aquatic Toxicity Space." In addition to K_{ow}, other molecular descriptors and chemical properties would probably be required for the delineation of subsequent QSARs. However, to predict acute and chronic toxicity reliably from any given set of structural parameters, the QSAR developed must first be identified with the proper mode of action.

11.3.3 Fish Acute Toxicity Syndromes and QSAR Development

Further implementation and refinement of QSAR in aquatic toxicology has required the development of qualitative SAR that assign the mode of action, and respective QSAR, to a given chemical structure. This effort involved creating a "toxic mode-of-action" database through the use of fish acute toxicity syndromes (FATS) as well as the definition and quantitation of the appropriate chemical properties (McKim et al. 1987c) used to classify chemicals.

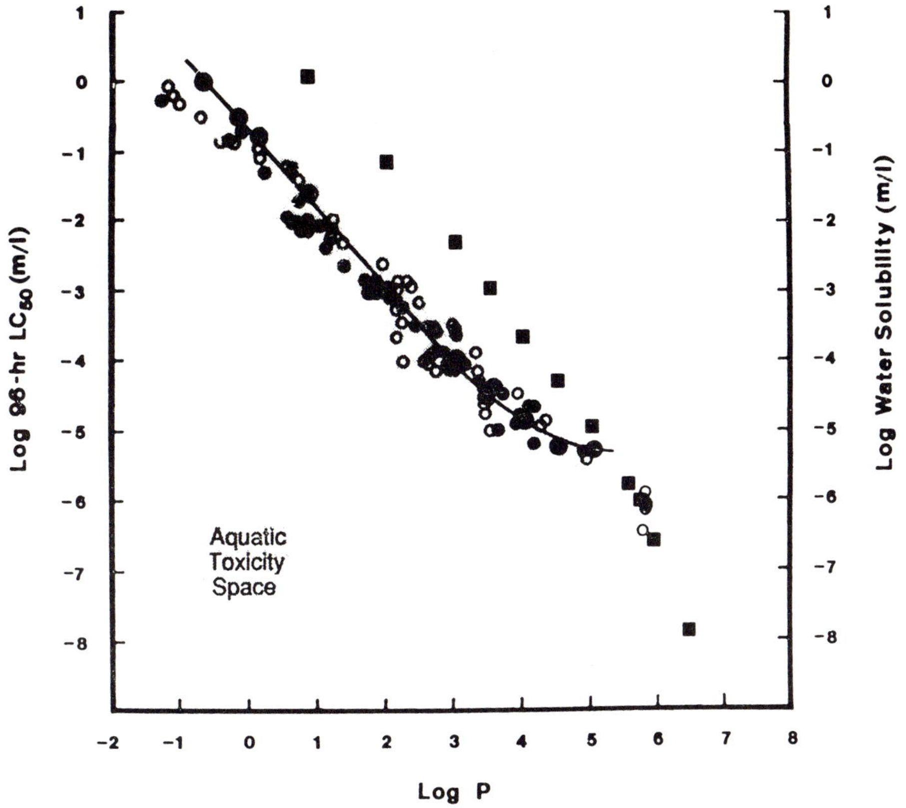

Figure 11.2 Relationship between log 96-h LC50 concentration with fathead minnows, Pimephales promelas, and 7-d LC50 with guppies, Poecelia reticulata, for narcotic chemicals. Fathead minnow data (solid circles) from Veith et al. (1983) and guppy data (open circles) from Konemann (1981). Solid squares represent water solubility data. The shaded area represents "aquatic toxicity space." (Modified from Veith et al. 1983).

The general protocol to build and use the FATS database in QSAR development is outlined in Figure 11.3. Through this experimental approach and statistical evaluation, investigators are successfully resolving a number of distinct FATS that correspond to specific modes of action (uncouplers, non-polar narcotics, AChE inhibitors, membrane irritants, respiratory blockers, polar narcotics, and central nervous system (CNS) seizure agents) (McKim et al. 1987a, b, c; Bradbury et al. 1989; Bradbury et al. 1991). These newly identified modes of action represent chemicals with toxicities greater than baseline (non-polar) narcosis. The lines described by the equations for these QSARs lie below the non-polar narcosis line as shown in Figure 11.4.

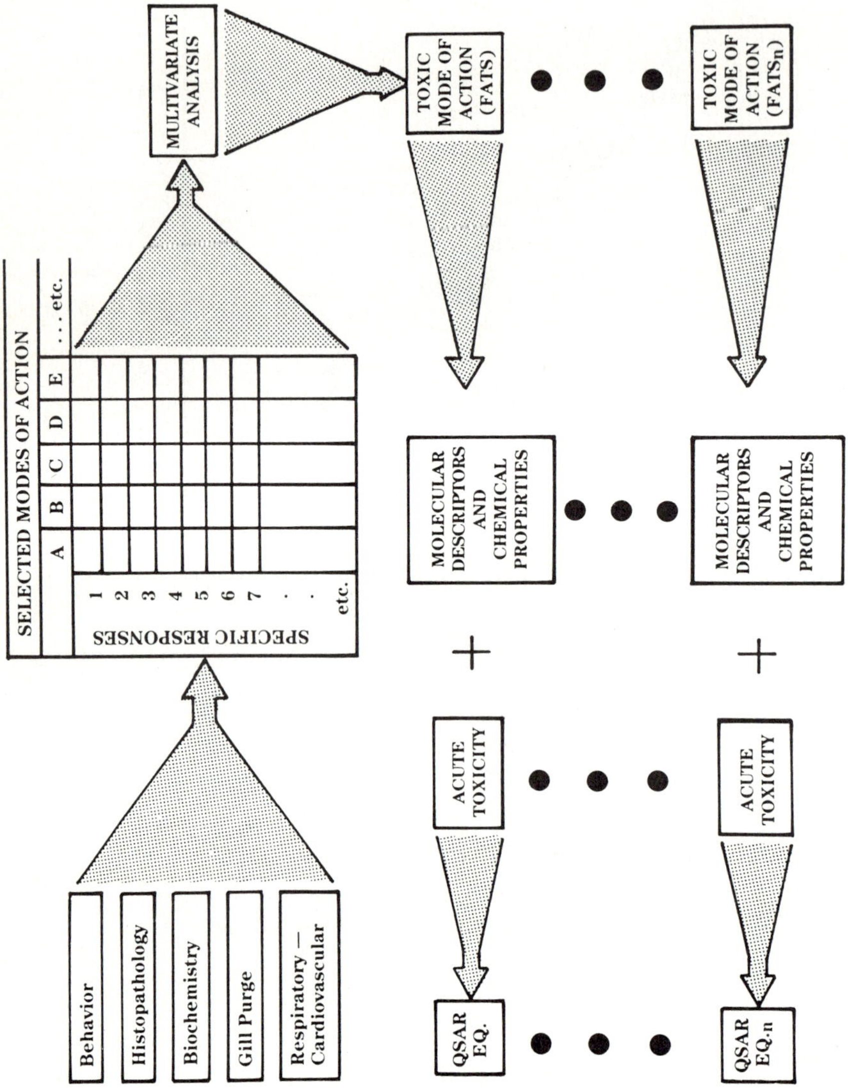

Figure 11.3 Flow chart relating fish acute toxicity syndrome (FATS) research to the generation of QSAR equations for predicting the toxicity of organic chemicals to fish. (From McKim et al. 1987).

Table 11.1 Some QSAR equations for nonspecific and unreactive industrial organic chemicals that define the nonpolar narcosis (baseline) mode of action in fish (modified from Hermens 1989)

Toxic Endpoint[a]	Type of Chemicals[b]	Equation[c] Log C (mol/l)	n	r^2	Reference
14-day LC50 to guppy	Chlorinated alkanes and aromatics, alcohols and ethers	-0.871 log P - 1.13	50	0.976	Konemann (1981)
96-h LC50 to fathead minnow	Alcohols, ketones, ethers, halogenated alkanes and aromatics	-0.94 log P + 0.94 log (0.000068 P + 1)-1.25	65	-	Veith et al. (1983)
96-h LC50 to goldfish	Alcohols and ketones	-1.0 log P - 0.8	8	0.97	Lipnick et al. (1987)
96-h LC50 to fish	Chlorobenzenes	-0.71 log P - 2.2	13	0.70	McCarty et al. (1985)
32-d MATC to fathead minnow	Chlorinated alkanes and aromatics, ketones and ethers	-0.886 log P - 2.18	10	0.925	Call et al. (1985)
Chronic toxicity to fish	Chlorobenzenes	-0.99 log P - 1.8	12	0.79	McCarty et al. (1985)

[a]LC50-Water concentration resulting in 50% mortality; MATC - maximum acceptable toxicant concentration, derived from early life stage test (ELS).

[b]The class of chlorinated alkanes and aromatics is restricted to unreactive representatives.

[c]C-Effect water concentration; P-octanol-water partition coefficient; n-number of chemicals in dataset; r^2-correlation coefficient.

QSARs for chemicals that are more specific in their activity than non-polar narcotics were developed for polar narcotics and fish (Table 11.2) and for uncouplers of oxidative phosphorylation and fish (Table 11.3). All of these QSARs show octanol/water partitioning to be the major controller of toxicity with other chemical descriptors such as pK_a having little or no significant impact on the QSAR produced. The slopes of the lines described by these two QSARs are slightly flatter than baseline narcosis indicating less of an effect of K_{ow} on the toxicity of polar narcotics and uncouplers than observed with non-polar narcotics (Figure 11.4). Researchers further suggest that the toxicity of highly specific reactive chemicals will be much less affected by log K_{ow} with even flatter QSARs residing below those shown in Figure 11.4 (Broderius et al. 1989).

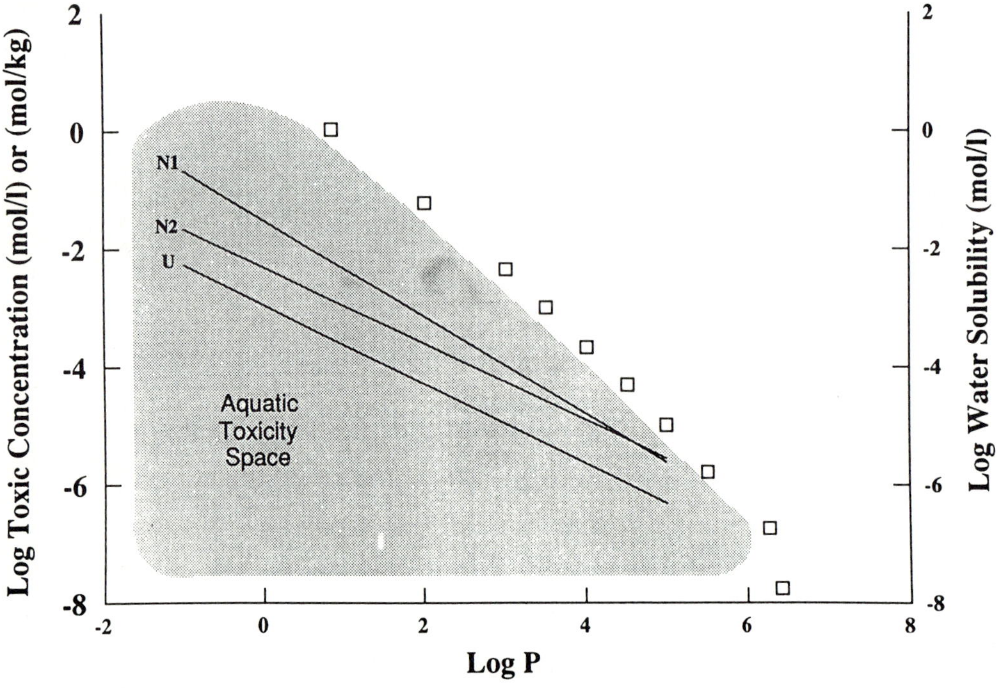

Figure 11.4 Relationship between water solubility (squares) and 96-hr LC50 toxicity QSARs for non-polar narcotics, N1; polar narcotics, N2; and uncouplers of oxidative phosphorylation across a wide range of log K_{ow}. The shaded area represents aquatic toxicity space (modified from Broderius et al. 1989).

Table 11.2 Some QSAR equations for industrial organic chemicals that define the polar narcosis mode of toxic action in fish (modified from Hermens 1989)

Toxic Endpoint[a]	Type of Chemicals[b]	Equation[c] Log C (mol/1)	n	r^2	Reference
14-day LC50 to guppy (at pH of 7.3)	Chlorophenols	-0.58 log P - 2.8 -1.1 log P - 0.35 pKa + 1.4	11 11	0.91 0.96	Konemann and Musch (1981)
96-h LC50 to guppy (at pH of 7.0)	Substituted phenols Ibid without two dinitrophenols	-0.52 log P - 3.0 -0.38 log P - 0.16 DpKa - 3.1 -0.59 log P - 2.7 -0.56 log P - 0.03 DpKa - 2.7	21 21 19 19	0.64 0.81 0.95 0.96	Saarikoski and Viluksela (1981)
96-h LC50 to fathead minnow	Some phenols and anilines	-0.65 log P - 2.3	39	0.90	Veith and Broderius (1987)
96-h LC to fathead minnow50	Substituted phenols	-0.601 log P - 2.43	21	0.938	Schultz et al. (1986)
96-h LC50 to rainbow trout	Substituted phenols	-0.61 log P - 3.1	5	0.96	McCarty et al. (1985)
14-day LC50 to guppy	Chloroanilines	-0.92 log P - 2.28	11	0.895	Hermens et al. (1984c)
Chronic toxicity to fish	Substituted phenols	-1.0 log P - 3.2	5	0.95	McCarty et al. (1985)

[a]LC50 - Water concentration resulting in 50% mortality.

[b]C - Effect water concentration; P - octanol-water partition coefficient; pKa - acid dissociation coefficient; n - number of chemicals in data set; r^2 - correlation coefficient; DpKa - difference in pKa between substituted phenol and phenol.

Table 11.3 Some QSAR equations for industrial organic chemicals that define the protonophoric respiratory uncoupler of oxidative phosphorylation mode of action to fish (modified from Broderius et al. 1989)

Toxic Endpoint[a]	Type of Chemicals[b]	Equation[c] Log C (mol/l)	n	r^2	Reference
96-hr LC50 to Atlantic salmon	Alkyl dinitro phenols	-0.31 log P - 5.3	6	0.98	Zitko et al. (1976)
96-h LC50 to fathead minnow	Substituted anilines and phenols	-0.590 log P - 3.25	11	0.917	Cajina-Quezada and Schultz (1989)
96-h LC50 to fathead minnow	Some phenols and anilines	-0.590 log P - 3.22	6	0.956	Schultz et al. (1986)
96-h LC to fathead minnow	Phenolic uncouplers	-3.931 log P - 0.447	5	0.981	Call et al. (1989)
32-d MATC to fathead minnow	Phenolic uncouplers	-4.942 log P - 0.344	5	0.992	Call et al. (1989)

[a]LC50 - Water concentration resulting in 50% mortality; MATC - Maximum Acceptable Toxicant Concentration, derived from early life stage test (ELS).

[b]C - Effect water concentration; P - octanol-water partition coefficient; n - number of chemcials in data set; r^2 - correlation coefficient.

11.4 Relationship Between Toxicity and Bioconcentration

Toxicologists define a toxic dose as the concentation of a chemical at a receptor or target site in the animal that elicits a toxic response. It is assumed for most chemicals that the toxic dose at the target site is similar across species and that differences in LD50s or LC50s between species are related to physiological or biochemical differences that change the kinetics of absorption, distribution, biotransformation and elimination. If the kinetics are accounted for, yet differences still exist between species, then it is assumed that differences in target site sensitivity exist. Differences in animal response to chemicals are also related to environmental factors which alter the kinetics through making the chemical more or less bioavailable. Bioavailability in aquatic tests is altered by environmental factors such as temperature, dissolved oxygen, water pH and dissolved and suspended organic material (Sprague 1985). The impact of these environmental factors on the mechanisms controlling the movement of organic chemicals across the gills of fish, a major exposure route, was recently reviewed by McKim and Erickson (1991). Their review of the recent literature indicated that environmental pH and dissolved organic matter produced a considerable reduction in the bioavailability of both ionizable organics and high log K_{ow} (> 5) organics, respectively. Environmental factors coupled with species, laboratory, protocol, and investigator differences combine to make toxicity comparisons and regulatory application of aqueous exposures difficult at best.

The relationship between a number of key thermodynamic properties octanol/water partition coefficients, bioconcentration factors, water solubility, and sediment/water partition coefficients, and toxicity was demonstrated in QSAR studies by Friant and Henry (1985). They point out that the major factor controlling these relationships is the dominance of the hydrophobic component of the molecule. This is especially true for those chemicals that are passive or nonreactive in their modes of toxic action as represented by narcotic chemicals (Franks and Lieb 1984). For reactive chemicals with specific modes of toxic action there is a mixture of both active and passive toxicity and hydrophobicity does not completely dominate the mechanism of toxicity. Hydrophobicity not only affects chemical partitioning, but also toxic response (Friant and Henry 1985). From the bioconcentration/log K_{ow} and toxicity/log K_{ow} relationships described earlier in the QSAR discussion, it is clear that the concentration of a chemical in an organism must also be related to its toxicity. For a toxic response to occur in an aquatic animal a chemical must enter the organism, via inhalation, oral or dermal routes, and be transported to a target or receptor site. At steady-state the internal concentration of the chemical is proportional to the chemical concentration in the water. Therefore, since water concentrations of organic chemicals are directly related to measured toxic responses of aquatic animals, tissue concentrations must also be directly related to toxic responses.

11.5 Estimated Toxic Internal Tissue Residues

In an effort to expand the predictive capabilities of QSAR in aquatic toxicology McCarty (1986, 1987a, b) evaluated the use of QSARs to estimate the toxic internal concentrations of organic chemicals in addition to the commonly used estimates of toxic water concentrations. Using the known relationships between toxicity and log K_{ow} and between bioconcentration and log K_{ow} described earlier, he evaluated the combination of toxicity and bioconcentration QSARs into a model for predicting toxic internal residues (Figure 11.5). Through the use of a geometric mean functional regression procedure he recalculated a series of QSAR regression equations from the literature such as those shown in Tables 11.1, 11.2, and 11.3 that describe the acute and chronic toxicity of selected organics to several species of fish. He then combined these equations with a bioconcentration relationship originally developed by MacKay (1982). The bioconcentration relationship was modified by Halfon (1985) using the geome-

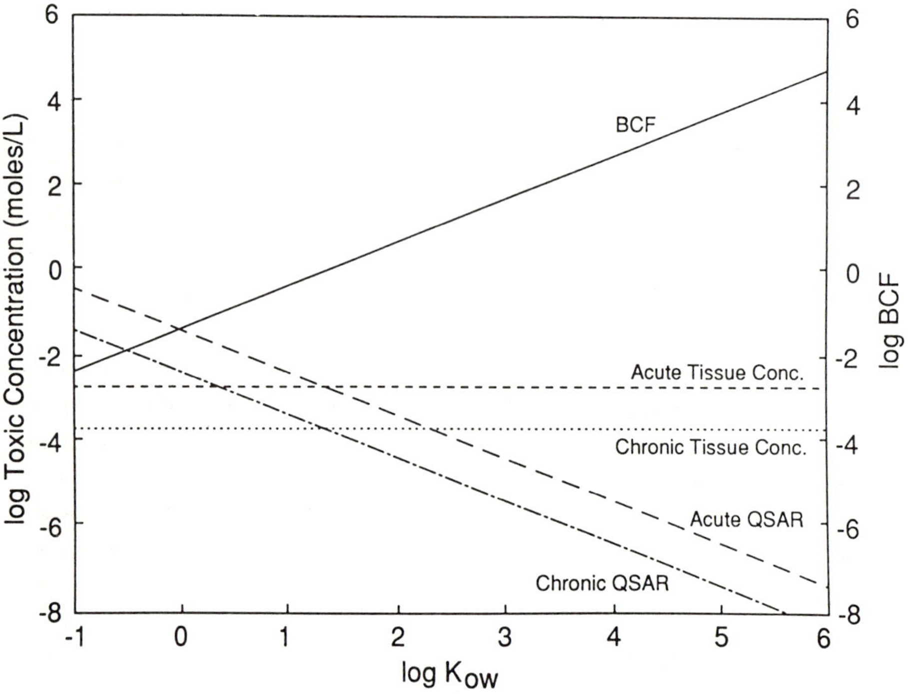

Figure 11.5 Relationship between toxicity, bioconcentration (log BCF) and octanol/water partition coefficient (log Kow) for some narcotic organic chemicals (modified from McCarty 1987a).

tric mean functional regression method, to produce a one constant equation (1):

$$\log BCF = 1.000 \log K_{ow} - 1.336 \qquad (1)$$

This modified equation was used as the bioconcentration factor equation in the final bioconcentration/toxicity relationship (McCarty 1986, 1987a). By substituting equation (1) of MacKay (1982) (bioconcentration/K_{ow} relationship) into the QSAR equations for fish acute and chronic non-polar narcosis toxicity (toxicity/K_{ow} relationship, equations 2 and 3), the internal acute and chronic tissue concentrations could be estimated for non-polar narcotics for many different external toxicant concentrations (McCarty 1986, 1987a). The toxicity/bioconcentration equations used by McCarty are as follows:

Logarithmic Form	Arithmetic Form	
Original Equations (McCarty 1987a)		
$\log BCF = \log K_{ow} - 1.336$	$BCF = 0.0461 * K_{ow}$	(1)
$\log acute = -\log K_{ow} - 1.38$	$acute = 1/K_{ow} * 0.042$	(2)
$\log chronic = -\log K_{ow} - 2.33$	$chronic = 1/K_{ow} * 0.0047$	(3)
Derived Equations (McCarty 1987a)		
$EQ\ 1+2 = \log acute = -\log BCF - 2.72$	$acute = 0.0019 .1/BCF$	(4)
$EQ\ 1+3 = \log chronic = -\log BCF - 3.67$	$chronic = 0.0002. 1/BCF$	(5)

The manipulation of toxicity and bioconcentration QSARs as shown above is necessary to develop the QSAR relationship; however, the estimation of internal toxic residue concentrations is done essentially by multiplying the calculated threshold LC50 value for each chemical by its respective calculated BCF (assuming 5% lipid content, $\log BCF = \log K_{ow} -1.3$) (McCarty 1987a) resulting in a toxicity/bioconcentration-based residue estimate (TBRE).

Calculating the whole-body residues associated with specific toxic endpoints provides a separation of the kinetics, controlled by hydrophobicity, from the toxicity, controlled by chemical concentration at the site of toxic action. Ferguson (1939) suggested that within a similar group of animals, and within a group of chemicals with the same mode of action, biological responses are associated with critical body residues which approximately equal chemical activity. The precise value of these body residues is controlled by the specific toxic endpoint being studied as observed for acute and chronic tissue concentrations in Figure 11.5. As discussed earlier, as the mode of action becomes more specific the toxicity QSAR and the toxic residues are shifted downward within "aquatic toxicity space" indicating increased toxicity.

McCarty's work on the derivation of TBRE is particularly important in the QSAR approach to predictive toxicology in that it shows that hydrophobicity has little predictive or interpretive value for toxicant action or chemical potency, although, it does describe the partitioning behavior of organic chemicals. When one compares the LC50-based QSAR and the TBRE in Figure 11.5, there is a range of 7 orders of magnitude for LC50-based QSAR toxicity estimates, while the TBREs are considered equally potent and range in value within one order of magnitude. Again, the LC50 based QSARs, which use aqueous chemical concentrations combine both kinetics (controlled by bioavailability) and potency information, while TBREs deal only with potency and toxicity information.

The mean estimated internal toxicant concentrations determined by McCarty (1987a) using a TBRE for fathead minnows exposed to non-polar narcotic chemicals ranged from 1.00 to 1.90 mmol/Kg for acute toxicity and from 0.20 to 0.40 mmol/Kg for chronic toxicity. Chronic TBRE for non-polar narcotics were approximately an order of magnitude lower than the acute values for chemicals within this mode of action. This agrees well with the difference between acute and chronic toxicity QSARs for non-polar narcotics (Call et al. 1985). TBREs for chemicals that were not narcotics ranged from 0.31 to 0.55 mmol/Kg acutely and 0.03 to 0.22 mmol/Kg chronic toxicity. The acute TBRE for substituted phenols were 2 to 4X lower than for non-polar narcotics while the chronic values were another order of magnitude lower than the acute residues (McCarty 1987a). The lower tissue residue estimates for the more specific-acting non-narcotic modes of action supports the idea that the non-polar narcotic mode of action makes up the baseline toxicity QSAR, and all other modes of action that are more specific and more toxic reside further down in "aquatic toxicity space."

A recent paper by Donkin et al. (1989) has also demonstrated that hydrophobicity influences toxicity mainly through its effect on bioconcentration. This was shown by expressing toxicity as the toxicant concentration in the soft tissue of a marine mussel (Mytilus edulis) that caused a 50% reduction in its feeding rate. For the non-polar narcotic compounds tested that had a log K_{ow} less than 4.6, the tissue concentrations causing a 50% reduction in feeding rate were constant (equally potent). Compounds with log K_{ow} values exceeding 5 were concentrated to a higher level in the tissues prior to affecting feeding rate, which indicated a possible 'molecular weight cutoff' in the toxic response. The interrelationships between hydrophobicity, sublethal toxicity and critical tissue residues observed by Donkin et al. (1989) are identical to those demonstrated by McCarty (1986) in Figure 11.5. However, in addition to estimating critical tissue concentrations, Donkin and co-workers actually measured them. Equal potency of these measured tissue concentrations lends further support to the proposed TBRE approach.

11.6 Comparison of Measured Internal Toxic Residues to Toxicity/Bioconcentration-Based Residue Estimates

As discussed earlier, the development of unique toxicity QSARs is feasible only if the QSAR represents a distinct mode of toxic action (McKim et al. 1987c). Therefore, it seems reasonable to assume that a distinct mode of action would also be accompanied by a specific critical body residue. At this time in aquatic toxicology distinct toxicity QSARs have been identified and developed for 1) acute and chronic non-polar narcotics (Veith 1983; Call 1985), 2) acute and chronic polar narcotics (Veith and Broderius 1987; Broderius et al. 1990), and 3) acute and chronic uncouplers of oxidative phosphorylation (Call et al. 1989; Cajina-Quezada and Schultz 1990). We have grouped existing body residue data (Table 11.4) that was available for chemicals falling within these three specific modes of action to evaluate the possible differences and/or similarities in critical body residues responsible for the acute and chronic toxic endpoints recorded. Also included is residue data for two chemicals assumed to be "CNS seizure agents," a newly defined FATS (Bradbury et al. 1991), that does not as yet have a defined QSAR. Using calculated BCFs we compared the measured lethal concentrations with the estimated lethal tissue concentrations (TBRE) calculated by multiplying the LC50 by the appropriate BCF (measured or calculated from log K_{OW}) (Table 11.4).

The measured mean lethal and sublethal residues for non-polar narcotics in guppies and mussels (Table 11.4) agreed quite well with the TBRE estimated values for lethal (1.0 to 1.9 mmol/Kg) and sublethal (0.2 to 0.4 mmol/Kg) non-polar narcotics determined by McCarty (1986, 1987a). Lethal and sublethal residues in Table 11.4 estimated from calculated BCFs were somewhat higher than measured residues. A major factor here is that calculated BCFs are considered steady-state and for toxicity tests of short duration with chemicals with log k_{OW} greater than 3 steady-state would not be attainable. Therefore, lethal residues using calculated BCFs would be somewhat overestimated as seen in Table 11.4. The agreement between measured toxic residues and estimated toxic residues for non-polar narcotics in Table 11.4 ranged from a factor of 2 to 4 for lethal residues and by a factor of 10 for sublethal endpoints.

Polar narcotic toxicity QSARs represented by the substituted anilines and phenols are listed in Table 11.2. These chemicals are known to be somewhat more toxic than non-polar narcotics (Veith and Broderius 1987) and McCarty's (1987a) TBRE ranged from 0.31 to 0.55 mmol/Kg, not quite an order of magnitude below his estimate for non-polar narcotics. The measured mean lethal residue for five chlorophenols and phenols (polar narcotics) in the goldfish (Table 11.4) was 1.03 + 0.353 mmol/Kg which was lower than the measured lethal residues in guppies for non-polar narcotics (Table 11.4). The estimates of lethal residues for these polar narcotics, using calculated BCFs, gave higher (5X) residues. No sublethal residues were available for the polar narcotics.

Table 11.4 Measured and estimated toxic endpoint residues in four aquatic species exposed to 28 waterborne organic chemicals representing four toxic mode of action

Toxic Mode of Action and Chemical Name	Molecular Weight (g)	Log Kow[1]	Animal Species[2]	Exposure Time (h)	Lethal Water Concentration (mmol/L)	Measured Residue at Toxic Endpoint (mmol/kg)	Calculated BCF[3]	Est. Lethal Residue LC_{50} x Calc. BCF (mmol/kg)	References
NON-POLAR NARCOTICS									
Lethal (LC_{50})									
1,3-Dichlorobenzene	147	3.57	FH	96	0.053	N/A	262	13.8	Carlson and Kosian (1987)
1,4-Dichlorobenzene	147	3.57	FH	96	0.029	N/A	262	7.60	Carlson and Kosian (1987)
1,2,4-Trichlorobenzene	181	4.28	FH	96	0.016	N/A	959	15.34	Carlson and Kosian (1987)
1,2,3,4-Tetrachlorobenzene	216	4.99	FH	96	0.005	N/A	3510	17.55	Carlson and Kosian (1987)
1,2,3-Trichlorobenzene	181	4.28	GP	2.4	0.056	2.71	959	53.7[4]	Van Hoogen and Opperhuizen (1988)
1,2,3-Trichlorobenzene			GP	24	0.004	2.02	959	3.83	Van Hoogen and Opperhuizen (1988)
1,2,3-Trichlorobenzene			GP	96	0.002	2.38	959	1.91	Van Hoogen and Opperhuizen (1988)
1,2,3,4-Tetrachlorobenzene	216	4.99	GP	96	0.0017	2.31	3510	5.97	Van Hoogen and Opperhuizen (1988)
1,2,3,4-Tetrachlorobenzene			GP	192	0.0011	2.64	3510	3.86	Van Hoogen and Opperhuizen (1988)
Pentachlorobenzene	250	5.71	GP	96	0.0005	2.54	12800	6.40	Van Hoogen and Opperhuizen (1988)
Pentachlorobenzene			GP	192	0.0004	2.11	12800	5.12	Van Hoogen and Opperhuizen (1988)
						2.39 ± 0.26(7)[5]		8.14 ± 5.43(10)	
Sublethal (EC_{50})									
Toluene	92	2.79	MU	1.3	0.0255	0.169	64	1.63	Donkin et al. (1989)
n-Propylbenzene	120	3.85	MU	1.3	0.0072	0.225	437	3.15	Donkin et al. (1989)
Naphthalene	128	3.32	MU	1.3	0.0072	0.245	166	1.20	Donkin et al. (1989)
Biphenyl	154	4.03	MU	1.3	0.0019	0.101	608	1.16	Donkin et al. (1989)
Acenaphthene	154	4.07	MU	1.3	0.0025	0.191	654	1.64	Donkin et al. (1989)
1-Chloronapthalene	163	4.03	MU	1.3	0.0017	0.133	607	1.03	Donkin et al. (1989)
Phenanthrene	178	4.49	MU	1.3	0.0008	0.173	1400	1.12	Donkin et al. (1989)
Dibenzothiophene	184	4.56	MU	1.3	0.0005	0.077	1580	0.79	Donkin et al. (1989)
n-Octane	114	4.93	MU	1.3	0.0011	0.216	3100	3.41	Donkin et al. (1989)
Fluoranthene	202	4.95	MU	1.3	0.0004	3.102[4]	3240	1.30	Donkin et al. (1989)
Pyrene	202	4.95	MU	1.3	0.0002	>0.936[4]	3240	0.65	Donkin et al. (1989)
						0.17 ± .057(9)		1.55 ± 0.91(11)	

Table 11.4 continued

Toxic Mode of Action and Chemical Name	Molecular Weight (g)	Log Kow[1]	Animal Species[2]	Exposure Time (h)	Lethal Water Concentration (mmol/L)	Measured Residue at Toxic Endpoint (mmol/kg)	Calculated BCF[3]	Est. Lethal Residue LC_{50} x Calc. BCF (mmol/kg)	References
POLAR NARCOTICS									
Lethal (LC_{50})									
Phenol	94	1.48	GF	24	0.638	1.213	5.82	3.71	Kobayashi et al. (1979)
2-Chlorophenol	128	2.20	GF	24	0.125	1.000	2.20	2.75	Kobayashi et al. (1979)
4-Chlorophenol	128	2.48	GF	24	0.070	0.789	36.6	2.56	Kobayashi et al. (1979)
2,4-Dichlorophenol	168	3.07	GF	24	0.046	1.595	105	4.83	Kobayashi et al. (1979)
2,4,6-Trichlorophenol	197	3.57	GF	24	0.051	1.015	265	13.51	Kobayashi et al. (1979)
2,4,5-Trichlorophenol	197	3.85	GF	24	0.009	0.569	441	3.97	Kobayashi et al. (1979)
						1.030 ± 0.353(6)		5.22 ± 4.15(6)	
UNCOUPLERS									
Lethal (LC_{50})									
2,3,4,6-Tetrachlorophenol	232	4.32	GF	24	0.003	0.323	1040	3.120	Kobayashi et al. (1979)
Pentachlorophenol	266	3.32	GF	24	0.001	0.357	167	0.167	Kobayashi et al. (1979)
Pentachlorophenol			FH	96	0.00098	0.282	167	0.163	Spehar et al. (1985)
Pentachlorophenol	266	3.32	FH	768	0.00061	0.282	167	0.102	Spehar et al. (1985)
						0.311 ± 0.036(4)		0.888 ± 1.49(4)	
SEIZURE AGENTS									
Lethal (LC_{50})									
trans-Permethrin	391.3	6.50	MD	48	0.000141	0.0097	54300	7.656	Kikuchi et al. (1984)
Fenvalerate	419.9	6.20	FH	45	0.000002	0.0029	31500	0.056	Bradbury et al. (1985)
						0.0063 ± .0048(2)		3.86 ± 5.37(2)	

[1]Calculated Log Kow using CLOGP. (Leo and Weininger, 1985).
[2]Fathead minnow = FH, Guppy = GP, Goldfish = GF, Mussel = MU, Medaka = MD.
[3]Calculated BCF (Veith and Kosian, 1982)
[4]Values were not included in mean calculation.
[5]Mean ± Std (N)

The most recent addition to toxicity QSAR is the acute and chronic QSAR for uncouplers of oxidative phosphorylation (Table 11.3). At the present time, no TBRE are available for uncouplers, however, toxic residues were measured in fish exposed to lethal concentrations of tetrachlorophenol and pentachlorophenol, two classic uncouplers (Table 11.4). The mean toxic residue for the uncoupler mode of action is about an order of magnitude lower than the mean non-polar narcosis lethal residues. This agrees well with the differences between the acute toxicity QSARsfor these two modes of action (Veith et al. 1983; Call et al. 1989).

In summary, the data on measured and estimated residues compiled in Table 11.4 and summarized in Figure 11.6 seem to support the TBRE of McCarty (1987a) for acute and chronic effects of non-polar narcotics. In addition, measured and estimated lethal residues for two new modes of action (polar narcosis and uncouplers) also lend support to the TBRE approach. The residues observed for these two new modes of action are lower than those related to non-polar narcosis with uncoupler residues being lower than polar narcosis and the pyrethroid insecticides, fenvalerate and permethrin, highly specific neurotoxins, showing the lowest toxic residues yet observed (Table 11.4, Figure 11.6). However, the super hydrophobicity of the seizure agents (log K_{ow} = 6.5) seems to eliminate the usefulness of estimating residues from calculated BCFs because of extremely high adipose fat storage as compared to the target organ concentrations (Donkin et al. 1989; Van Hoogen and Opperhuizen 1988). There seems to be a clear delineation in toxic residues between the 4 established modes of action (Figure 11.6). It is also interesting to note the similar potency of those chemicals plotted within each of the toxic modes of action. One can readily see the large differences in aqueous toxicity between chemicals within the same mode of action with varying log K_{ow}, while internal tissue concentrations remain similar across log K_{ow}.

The similarity noted in Figure 11.6 between the species within toxic modes tested was quite remarkable, since numerous factors were apparent that could modify residue levels in fish such as, species sensitivity, lipid content, investigator and laboratory differences. There were certainly differences in lipid content between the species used in these comparisons. Lipid content can be affected by a number of factors and varies widely (Henderson and Tocher 1987; Halver 1989). Fish with high lipid content would tend to have higher interior tissue concentrations than those with lower lipid concentrations yet no large variations in measured residue were observed here.

11.7 Internal Toxic Residues and Mode of Action

Through a combination of multivariate statistics and fish behavioral, physiological and biochemical studies, it was possible to develop a set of Fish Acute Toxicity Syndromes

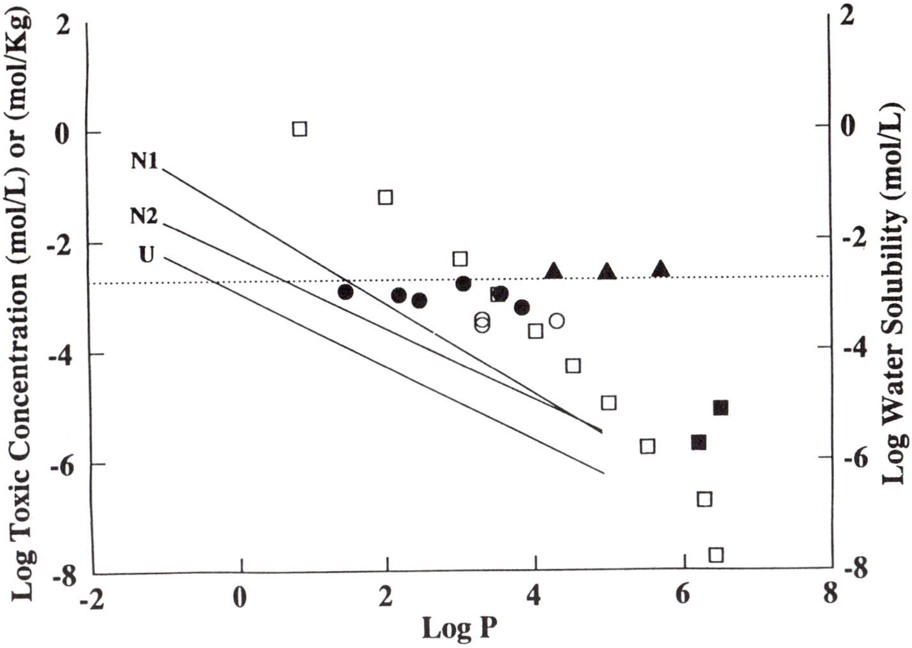

Figure 11.6 Water exposure concentration (LC50)/fish lethal residue concentration in relation to solubility and octanol/water partition coefficients. N1 = QSAR defining non-polar narcosis mode of action based on Fathead Minnow (FH) LC50s; N2 = QSAR defining polar narcotic mode of action based on FH LC50s; U = QSAR defining oxidative phosphorylation uncoupling mode of action based on FH LC50s (Broderius et al. 1989). .... = Estimated lethal residue concentration for non-polar narcotics (McCarty et al. 1986, 1987a); □ = Water solubility. ▲= Guppy lethal residues for non-polar narcotic chemicals (Van Hoogen and Opperhuizen 1988); ,● = Goldfish lethal residues for polar narcotic mode of action (Kobayashi et al. 1979); ○ = Goldfish and FH lethal residues for uncouplers (Kobayashi et al. 1979; Spehar et al. 1985); ■ Medaka and FH lethal residue concentrations for central nervous system seizure agent mode of action (Kikuchi et al. 1984; Bradbury et al. 1985).

(FATS) that accurately separated chemicals with distinctive structures into 7 separate modes of toxic action (McKim et al. 1987a, b, c; Bradbury et al. 1989; Drummond et al. 1986; Drummond et al. 1990; Bradbury et al. 1991). These seven modes of action included: 1) non-polar narcosis, 2) polar narcosis, 3) uncouplers of oxidative phosphorylation, 4) skin irritants, 5) AChE inhibitors, 6) respiratory blockers, and 7) Central Nervous System seizure agents. More recently, to get a first approximation of how lethal residues vary with different modes of action, Schmieder et al. (1991) using large rainbow trout in respirometer-metabolism chambers determined the total net dose absorbed at death (total internal residue at death) for 56 animals exposed to 14 chemicals representing 6 specific modes of action. As noted in Table 11.5, the non-polar narcotics which represent "baseline toxicity" have the highest lethal residues measured at death, with polar narcosis showing the next highest residues. Tissue concentrations of uncouplers at death were considerably lower than for both groups of narcotics which indicates a more toxic and specific mode of action. AChE inhibitors demonstrated tissue residues at death which also indicated a more specific mode of action than simple narcosis yet similar in magnitude to uncouplers. The most specific mode of action observed with these trout dealt with the respiratory blocker, rotenone, which was several orders of magnitude more toxic based on the lethal residues in Table 11.5 than any other chemical tested. Since the skin irritants were surface acting at the route of entry across the gills, little confidence can be placed on their consistent residue accumulation. This is borne out by the wide variations between acrolein and benzaldehyde residues observed. A systemic dose is not important to the toxicity of these surface reactive chemicals.

Lethal residues for all chemicals within each mode of action group, Table 11.5, seem to agree. They were all similar in potency as observed in Figure 11.7, and the lethal residues determined for each mode of action were distinctly different from the others. There was, however, a slight downward trend in lethal tissue concentrations as log K_{ow} increased. In some cases the exposure regimes were very short, and for the more hydrophobic chemicals this would conceivably cause the target site tissue to reach a toxic level before the whole body concentrations. The internal distribution system is overwhelmed and can not distribute chemicals fast enough to the fat storage sites to bring all tissues to steady-state. Still, the values observed in Figure 11.7 are very close and show similar critical body residues for different exposure regimes, body size, and lipid content.

Overall, it would appear that chemicals within each of the modes of action have similar potency both for the trout study, Table 11.5, and for the other 5 species summarized from the literature (Table 11.4). There also seems to be a critical tissue residue established both for mode of action and for the toxic endpoints selected which tends to hold up across all of the studies reviewed.

Further work is needed on critical chemical residues measured in relation to specific toxic endpoints, exposure duration, and selected modes of toxic action. This would support the development of a more accurate predictive capability for TBRE. The present predictive capability is designed primarily for non-polar narcotics and could vary greatly when chemicals with more specific modes of action are encountered.

Table 11.5 Lethal residues in fifty-four large rainbow trout (600-1000 g) exposed to acutely lethal water concentrations of 14 organic chemicals representing 6 toxic modes of action (modified from Schmieder et al. 1991)

	Molecular Weight (g)	Log Kow[1]	Exposure Time (h)	Lethal Water Conc (mmol/L)	Lethal Residue[2] Conc (mmol/Kg)	Calculated BCF[3]	Est. Lethal Residue Water Conc x Calc BCF (mmol/Kg)	Fish (N)
NON-POLAR NARCOSIS								
MS222	261.3	1.96	6.8 ± 0.4	.195 ± 0.021	1.71 ± 0.23	14.1	2.75 ± 0.30	2
Octanol	130.2	2.94	5.7 ± 3.6	.118 ± 0.004	1.64 ± 1.25	83.5	9.89 ± 0.36	4
Overall Mean			6.3 ± 0.8	0.157 ± 0.054	1.68 ± 0.05		6.32 ± 5.05	
POLAR NARCOSIS								
Aniline	93.1	0.915	3.3 ± 1.7	0.736 ± 0.024	0.96 ± 0.59	2.1	1.55 ± 0.050	4
Phenol	94.1	1.48	13.4 ± 6.0	0.094 ± 0.002	0.78 ± 0.49	5.8	0.55 ± 0.011	4
2-Chloroaniline	127.5	1.90	17.2 ± 8.1	0.0135 ± 0.001	0.23 ± 0.15	12.6	0.17 ± 0.007	4
4-Chloroaniline	127.5	1.83	5.8 ± 3.2	0.175 ± 0.004	0.93 ± 0.86	11.1	1.94 ± 0.041	4
2,4-Dimethylphenol	122.2	2.77	5.9 ± 1.5	0.075 ± 0.002	0.49 ± 0.07	61.8	4.60 ± 0.147	4
Overall Mean			9.1 ± 5.9	0.219 ± 0.295	0.68 ± 0.31		1.76 ± 1.74	
UNCOUPLERS								
2,4-Dinitrophenol	184.1	1.91	15.0 ± 8.1	0.026 ± 0.002	0.160 ± 0.08	13.0	0.332 ± 0.02	4
Pentachlorophenol	266.3	3.32[4]	32.0 ± 4.7	0.00035 ± 0.00004	0.052 ± 0.03	170.	0.060 ± 0.07	4
Overall Mean			23.5 ± 12.0	0.013 ± 0.018	0.106 ± 0.08		0.196 ± 0.19	
ACHE INHIBITORS								
Malathion	330.4	1.76	40.3 ± 12.0	0.00091 ± 0.00017	0.149 ± 0.08	9.7	0.009 ± 0.002	4
Carbaryl	201.2	2.39	13.8 ± 8.9	0.025 ± 0.0015	0.178 ± 0.05	30.5	0.755 ± 0.046	4
Overall Mean			27.0 ± 18.7	0.013 ± 0.017	0.164 ± 0.02		0.382 ± 0.528	
RESPIRATORY BLOCKER								
Rotenone	394.4	3.49	20.7 ± 8.6	0.00001 ± 0.000001	0.0009 ± 0.0002	228.0	0.0028 ± 0.0003	4
Overall Mean			20.7	0.00001	0.0009		0.0028	
RESPIRATORY IRRITANTS								
Acrolein	56.1	0.101	20.5 ± 5.6	.0014 ± 0.0001	0.094 ± 0.035	1.0	0.0014 ± 0.0001	4
Benzaldehyde	106.1	1.50	19.3 ± 1.1	.357 ± 0.0057	13.22 ± 7.46	6.0	2.14 ± 0.034	4
Overall Mean			19.9 ± 0.9	0.179 ± 0.251	6.66 ± 9.28		1.07 ± 1.14	

[1]Calculated Log Kow using CLOGP. (Leo and Weininger, 1985).

[2]Calculated from net extraction efficiency data and water concentration from fish in metabolism chambers.

[3]Calculated BCF (Veith and Kosian, 1982).

[4]Apparent Log P for Water pH of 8.0 (Kaiser and Valdmanis, 1982).

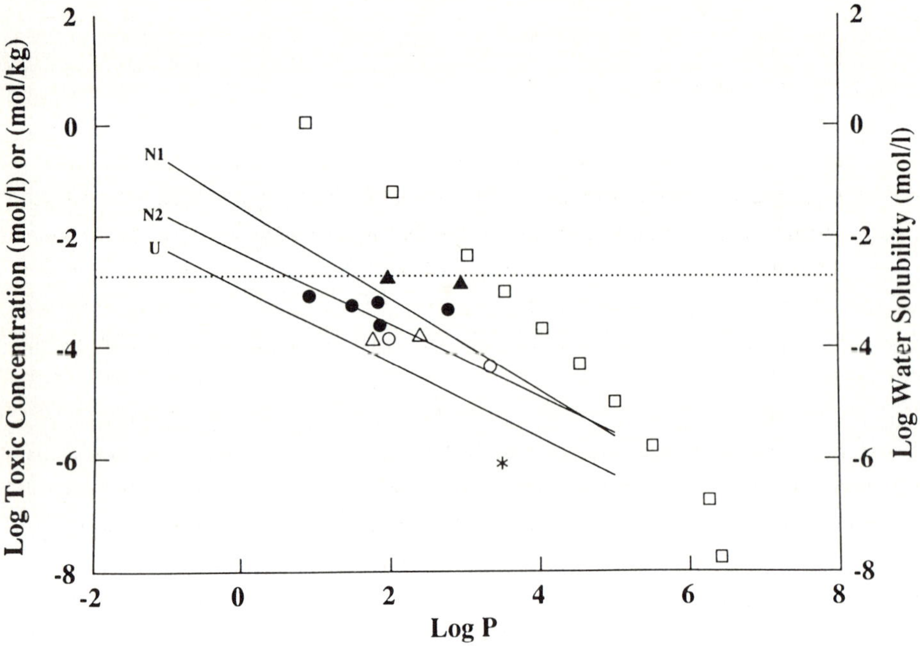

Figure 11.7 Water exposure concentration (LC50)/Rainbow Trout (RBT) lethal residue concentration in relation to water solubility and octanol/water partition coefficients. N1 = QSAR defining non-polar narcosis mode of action based on Fathead Minnow (FH) LC50s; N2 = QSAR defining polar narcotic mode of action based on FH LC50s; U = QSAR defining oxidative phosphorylation uncoupling mode of action based on FH LC50s (Broderius et al. 1989). .... = Estimated lethal residue concentration for non-polar narcotics (McCarty et al. 1986, 1987a); □ = Water Solubility. ▲ = RBT lethal residues for non-polar narcotic chemicals; ● = RBT lethal residues for polar narcotic chemicals; ○ = RBT lethal residues for uncoupler chemicals; * = Lethal residues for respiratory blocker chemicals. (Modified from Schmieder et al. 1991).

Finally, for those modes of action identified so far by toxicity QSARs the acute and chronic toxicity regressions are parallel. This indicates that the sensitivity and perhaps the accuracy of critical residue estimates could be enhanced by incorporating sublethal or chronic endpoints into these studies.

11.8 Conclusions

As we have tried to emphasize in this review, bioaccumulation is not reflective of toxicity. It merely informs us of the activity of the chemical and how it will act kinetically. Whether or not a chemical is highly bioaccumulatable says nothing about its toxic potency. By combining bioaccumulation and toxicity relationships as described here it is possible to remove the kinetic aspects involved in the use of aqueous chemical concentrations in toxicity relationships, and concentrate on the true potency of a chemical or group of chemicals with a similar toxic mode of action.

TBRE have provided good first approximation estimates of toxic internal residues for those chemicals which have a non-specific toxic mode of action such as non-polar narcosis. Further studies with industrial chemicals that represent a more specific toxic mode of action than non-polar narcosis demonstrate that TBRE designed specifically for each new mode of action may also be the most effective first approximation of toxic residues. Since each mode of action identified has shown a characteristic lethal residue it seems imperative that TBRE be developed, as are toxicity QSARs, for chemicals representing specific toxic modes of action.

When comparing measured lethal residues in fish to those estimated using the LC50 and the calculated BCF (TBRE) the estimates are usually higher than the measured values. Factors which may account for this variation are: 1) differences in lipid content of the animals, 2) variations in species target organ sensitivities, 3) calculated BCFs are steady-state and for many chemicals above log k_{ow} 3 steady-stateis not reached in short duration toxicity tests, and 4) the fact that calculated BCFs do not consider the impact of metabolism in their relationships. Furthermore, it is necessary to gain more experience in working with the lethal residues of chemicals with more specific toxic modes of action than narcosis so that researchers can better understand the processes and interrelationships involved with internal residues and specific toxicological endpoints.

More in-depth studies are needed to investigate: 1) the impact of duration of exposure on lethal residues or thresholds, 2) the differences in threshold residues associated with toxic endpoints other than lethality, 3) the effect of high log K_{ow} on the usefulness of total body residue as a surrogate for target organ residues, 4) the consistency and magnitude of the variations in toxic residues observed for different modes of action, 5) the importance of species sensitivity in the use of toxic body residues, 6) the accuracy of estimating toxic residues for mixtures of chemicals with the same mode of action, and 7) the proper use of a toxic residue approach in regulatory decisions.

11.9 References

Bradbury, S.P., Carlson, R.W., Niemi, G.J. and T.R. Henry (1991). Use of respiratory-cardiovasacular responses of rainbow trout (Oncorhynchus mykiss) in identifying acute toxicity syndromes in fish: Part 4. Central nervous system seizure agents. *Environ. Toxicol. Chem.*, 10, 115131.

Bradbury, S.P., Coats, J.R. and J.M. McKim (1985). Differential toxicity and uptake of two fenvalerate formulations in fathead minnows (Pimephales promelas). *Environ. Toxicol. Chem.*, 4, 533-541.

Bradbury, S.P., Henry, T.R., Niemi, G.J., Carlson, R.W. and V.M. Snarski (1989). Use of respiratory-cardiovascular responses of rainbow trout (Salmo gairdneri) in identifying acute toxicity syndromes in fish: Part 3. Polar narcotics. *Environ. Toxicol. Chem.*, 8, 247-261.

Broderius, S.J., Russom, C.L. and M. Nendza (1989). Mode of action-specific QSAR models for predicting acute and chronic toxicity of industrial organic chemicals to aquatic organisms. Deliverable No. 8142A, U.S. Environmental Protection Agency, Environmental Research Laboratory, Duluth, MN 55804. 110 pp.

Cajina-Quezada, M. and T.W. Schultz (1990). Structure-toxicity relationships for selected weak acid respiratory uncouplers. *Aquat. Toxicol.*, 17, 239-252.

Calamari, D., Galassi,S., Setti, F. and M. Vighi (1983). Toxicity of selected chlorobenzenes to aquatic organisms. *Chemosphere*, 12, 253-262.

Call, D.J., Brooke,L.T., Knuth, M.L., Poirier, S.H. and M.D. Hoglund (1985). Fish subchronic toxicity prediction model for industrial organic chemicals that produce narcosis. *Environ. Toxicol. Chem.*, 4, 335-341.

Call, D.J., Poirier,S.H., Lindberg, C.A., Harting,S.L., Markee,T.P., Brooke, L.T., Zarvan, N. and C.E. Northcott (1989). Toxicity of selected uncoupling and acetylcholinesterase-inhibiting pesticides to the fathead minnow (Pimephales promelas). In: D.L. Weigmann (ed.). Pesticides in Terrestrial and Aquatic Environments. Proceedings of a National Research Conference, May 11-12, 1989.

Carlson, A.R. and P.A. Kosian (1987). Toxicity of chlorinated benzenes to fathead minnows Pimephales promelas. *Arch. Environ. Contam. Toxicol.*, 16, 129-135.

Donkin, P., Widdows, J., Evans, S.V., Worrall, C.M. and M. Carr (1989). Quantitative structure-activity relationships for the effect of hydrophobic organic chemicals on rate of feeding by mussels Mytilus edulis. *Aquat. Toxicol.*, 14, 277-294.

Drummond, R.A. and C.L. Russom (1990). Behavioral toxicity syndromes: A promising tool for assessing toxicity mechanisms in juvenile fathead minnows. *Environ. Toxicol. Chem.*, 9, 37-46.

Drummond, R.A., Russom, C.L, Geiger, D.L. and D.L. DeFoe. (1986). Behavioural and morphological changes in fathead minnows (Pimephales promelas) as diagnostic endpoints ..for screening chemicals according to mode of action. In: T.M. Poston and R. Purdy (eds.). Aquatic Toxicology and Environmental Fate: Ninth Symposium. American Society of Testing and Materials, ASTM STP 921, Philadelphia, PA. pp. 415-435.

Ferguson, J. (1939). The use of chemical potentials as indices of toxicity. *Proc. Roy. Soc. Br.*, 127, 387-404.

Franks, N.P. and W.R. Lieb (1984). Do general anaesthetics act by competitive binding to specific receptors? *Nature*, 310, 599-601.

Friant, S.L. and L. Henry (1985). Relationship between toxicity of certain organic compounds and their concentrations in tissues of aquatic organisms: A perspective. *Chemosphere*, 14, 1897-1907.

Halfon, E. (1985). Regression method in ecotoxicology: A better formulation using the geometric mean functional regression. *Environ. Sci. Technol.*, 19, 747-749.

Halver, J.E. (1989). Fish nutrition. Second Edition. Academic Press, San Diego, CA. 798 pp.

Henderson, R.J. and D.R. Tocher (1987). The lipid composition and biochemistry of freshwater fish. *Prog. Lipid Res.*, 26, 281-347.

Hermens, J., Canton, H., Janssen, P. and R. deJong (1984b). Quantitative structure-activity relationships and toxicity studies of mixtures of chemicals with anaesthetic potency: Acute lethal and sublethal toxicity to Daphnia magna. *Aquat. Toxicol.*, 5,143-154.

Hermens, J., Canton, H., Steyger, N. and R. Wegman (1984a). Joint effects of a mixture of 14 chemicals on mortality and inhibition of reproduction of Daphnia magna. *Aquat. Toxicol.*, 5, 315-322.

Hermens, J., Leeuwangh, P. and A. Musch (1984). Quantitative structure-activity relationships and mixture toxicity studies of chloro- and alkylanilines at an acute toxicity level to the guppy (Poecilia reticulata). *Ecotoxicol. Environ. Saf.*, 8, 388-394.

Hermens, J.L.M. (1989). Quantitative structure-activity relationships of environmental pollutants. In: O. Hutzinger (ed.). Handbook of Environmental Chemistry. Vol. 2. Reactions and Processes. Springer-Verlag, Berlin.

Kikuchi, T., Nagashima, Y. and M. Chiba (1984). Accumulation and excretion of permethrin by the himedaka Oryzias latipes and biological significance of accumulated permethrin. *Bull. Jpn. Soc. Sci. Fish.*, 50, 101-106.

Kobayashi, K., Akitake, A. and K. Manabe (1979). Relation between toxicity and accumulation of various chlorophenols in goldfish. *Bull. Jap. Soc. Sci. Fish.*, 45, 173-175.

Konemann, H. (1981). Quantitative structure-activity relationships in fish toxicity studies. Part 1. Relationship for 50 industrial pollutants. *Toxicology*, 19, 209-221.

Konemann, H. and A. Musch (1981). Quantitative structure-activity relationships in fish toxicity studies. Part 2: The influence of pH on the QSAR of chlorophenols. *Toxicology* 19:223-228.

Leo, A. and D. Weininger (1985). Medchem Software Release 3.4. Medicinal Chemistry Project, Pomona College, Claremont, CA.

Lipnick, R.L., Pritzker, C.S. and D.L. Bentley (1987). Application of QSAR to model the acute toxicity of industrial organic chemicals to mammals and fish. In: Hadzi and Herman-Blazic (eds.). Progress and QSAR.

Mackay, D. (1982). Correlation of bioconcentration factors. *Environ. Sci. Technol.*, 16, 274-278.

McCarty, L.S. (1986). The relationship between aquatic toxicity QSARs and bioconcentration for some organic chemicals. *Environ. Toxicol. Chem.*, 5, 1071-1080.

McCarty, L.S. (1987a). Relationship between toxicity and bioconcentration for some organic chemicals. I. Examination of the relationship. In: K.L.E. Kaiser (ed.). QSAR in Environmental Toxicology - II. D. Reidel Publishing Company, Dordrecht, Holland. pp. 207-220.

McCarty, L.S. (1987b). Relationship between toxicity and bioconcentration for some organic chemicals. II. Application of the relationship. In: K.L.E. Kaiser (ed.). QSAR in Environmental Toxicology - II. D. Reidel Publishing Company, Dordrecht, Holland. pp. 221-230.

McCarty, L.S., Hodson, P.V., Craig, G.R. and K.L.E. Kaiser (1985). On the use of quantitative structure-activity relationships to predict the acute and chronic toxicity of chemicals to fish. *Environ. Toxicol. Chem.*, 4, 595-606.

McKim, J.M. and R.J. Erickson (1991). Environmental impacts on the physiological mechanisms controlling xenobiotic transfer across fish gills. *Physiol. Zool.* (In press).

McKim, J.M., Schmieder, P.K., Carlson, R.W., Hunt, E.P. and G.J. Niemi (1987a). Use of respiratory-cardiovascular responses of rainbow trout Salmo gairdneri in identifying acute toxicity syndromes in fish: Part 1. Pentachlorophenol, 2,4-dinitrophenol, tricaine methanesulfonate and 1-octanol. *Environ. Toxicol. Chem.*, 6, 295-312.

McKim, J.M., Schmieder,P.K., Niemi, G.J., Carlson, R.W. and T.R. Henry (1987b). Use of respiratory-cardiovascular responses of rainbow trout Salmo gairdneri in identifying acute toxicity syndromes in fish: Part 2. Malathion, carbaryl, acrolein and benzaldehyde. *Environ. Toxicol. Chem.*, 6, 313-328.

McKim, J.M., Bradbury, S.P. and G.J. Niemi (1987c). Fish acute toxicity syndromes and their use in the QSAR approach to hazard assessment. *Environ. Health Perspect.*, 71, 171-186.

McLeese, D.W., Zitko, V. and M.R. Peterson (1979). Structure-lethality relationships for phenols, anilines, and other aromatic compounds in shrimp and clams. *Chemosphere*, 2, 53-57.

Ribo, J.M. and K.L.E. Kaiser (1983). Effects of selected chemicals to photoluminescent bacteria and their correlations with acute and sublethal effects on other organisms. *Chemosphere*, 12, 1421-1442.

Rogerson, A., Shiu, W.Y., Huang, G.L., MacKay, D. and J. Berger (1983). Determination and interpretation of hydrocarbon toxicity to ciliate protozoa. *Aquat. Toxicol.*, 3, 215-228.

Saarikoski, J. and M. Viluksela (1981). Influence of pH on the toxicity of substituted phenols to fish. *Arch. Environ. Contam. Toxicol.*, 10, 747-753.

Schmieder, P.K., McKim, J.M., Kosian, P. and A. Hoffman (1991). Lethal residues in rainbow trout (Oncorhynchus mykiss) as related to fish acute toxicity syndromes. (Submitted to *Chemosphere*).

Schultz, T,.W., Holcombe, G.W. and G.L. Phipps (1986). Relationships of quantitative structure-activity to comparative toxicity of selected phenols in the Pimephales promelas and Tetrahymena pyriformis test systems. *Ecotoxicol. Environ. Saf.*, 12, 146-153.

Spehar, R.L., Nelson, H.P., Swanson, M.J. and J.W. Renoos (1985). Pentachlorophenol toxicity to amphipods and fathead minnows at different test pH. *Environ. Toxicol. Chem.*, 4, 389-398.

Sprague, J.B. (1985). Factors that modify toxicity. In: G.M. Rand and S.R. Petrocelli (eds.). Aquatic Toxicology. Hemisphere Publishing Corporation, New York, NY. pp. 124-163.

Van Hoogen, G. and A. Opperhuizen (1988). Toxicokinetics of chlorobenzenes in fish. *Environ. Toxicol. Chem.*, 7, 213-219.

Van Gestel, C.A.M. and W-C. Ma (1988). Toxicity and bioaccmulation of chlorophenols in earthworms, in relation to bioavailability in soil. *Ecotoxicol. Environ. Saf.*, 15, 289-297.

Veith, G.D. and S.J. Broderius (1987). Structure-toxicity relationships for industrial chemicals causing type (II) narcosis syndrome. In: K.L.E. Kaiser (ed.). QSAR in Environmental Toxicology - II. D. Reidel Publishing Company, Dordrecht, Holland. pp. 385-392.

Veith, G.D. and P. Kosian (1982). Estimating bioconcentration potential from octanol/water partition coefficients. In: MacKay, Paterson, Eisenreich and Simons (ed.). Physical Behavior of PCBs in the Great Lakes, Ann Arbor Science, Ann Arbor, chap. 15

Veith, G., Call, D. and L. Brooke (1983). Structure-toxicity relationship for the fathead minnow Pimephales promelas: Narcotic industrial chemicals. *Can. J. Fish. Aquat. Sci.* , 40, :743-748.

Veith, G.D., DeFoe, D.L. and B.V. Bergstedt (1979). Measuring and estimating the bio-concentration factor of chemicals in fish. *J. Fish. Res. Board Can.*, 36, 1040-1048.

Veith, G.D., DeFoe, D.L. and M. Knuth (1985). Structure-activity relationships for screening organic chemicals for potential ecotoxicity effects. *Drug Metab. Rev.*, 15, 1295-1303.

Zitko, V., McLeese, D.W., Carson, W.G. and H.E. Welch (1976). Toxicity of alkyldi-nitrophenols to some aquatic organisms. *Bull. Environ. Contam. Toxicol.*, 16, 508-515.

The Assessment of Bioaccumulation

Preben Kristensen and Henrik Tyle

12.1 Introduction

Knowledge on the actual fate of xenobiotic chemicals and heavy metals can only be achieved by profound chemical analysis of all compartments of the ecosystem. Due to cost and time limitations, such programmes have been restricted to substances which have shown profound effects in the environment (and to man), e.g. certain pesticides, PCB and a few heavy metals.

As more than 50,000 chemicals are commercially available and new chemicals are introduced every day, methods for screening the potential of hazard for man and the environment are necessary to prevent new hazardous chemicals, like DDT and PCB, to be released to the environment. Further it is necessary to regulate the use of certain dangerous new and existing substances to minimize environmental exposure.

Environmental hazard assessment is a tool for the evaluation of the potential threat to the environment of manufactured or marketed chemicals. The assessment is based on information on production, use and on intrinsic properties determined by the use of standardized methods regarding environmental fate and effects.

The potential for bioaccumulation is an important element in this assessment. At this stage no account is taken of specific local environmental conditions, e.g. bioaccumulation in various types of organisms or the environmental bioavailability of chemicals; elements which of course are highly important in more comprehensive environmental hazard and risk assessments.

Below some of the terms, which are used in this paper, will be defined.

Environmental Hazard Identification (EHI): The purpose of **EHI** is to identify substances of environmental concern. The **EHI** includes simple key parameters concerning intrinsic hazardous properties of the substances relevant for the potential environmental fate and effects (EEC 1990a, Lundgren 1989, Bro-Rasmussen 1986, 1988). **EHI** is used for selection of chemicals to be evaluated further or in special cases for regulatory purposes (OECD 1984a, 1986).

Environmental Hazard Assessment (EHA): The purpose of **EHA** is to assess the potential for a substance to cause adverse effects on free living species, if possible including populations, communities and ecosystems. Therefore, **EHA** includes a description of standard scenarios for which an exposure assessment, and to which the effects of the substance, can be related. Performing an **EHA** requires knowledge of:

- the environmental exposure (environmental compartments of concern, released quantities, i.e. rate of release and environmental fate and number of discharge points), and

- effect data on organisms with reference to the environmental compartment of concern.

EHA is often expressed by comparison of predicted or measured environmental concentrations (PECs and MECs, respectively) with the predicted no effect concentration (NEC) for the concerned species, populations, communities or ecosystems. When PEC or MEC is not available, or when these are very crude estimates, only an Initial Environmental Hazard Assessment can be made.

The comparison of PEC or MEC with NEC for the ecosystem might be performed in a variaty of ways, i.e. by the use of an uncertainty factor or application factor to "translate" acute toxicity data to the presumed no effect concentration of the ecosystem (Sloof et al. 1985, Bro-Rasmussen 1988, EEC 1990a, OECD 1989a, Hart and Jensen 1990).

Environmental Risk Assessment (ERA): The purpose of **ERA** is to determine the probability (qualitatively or quantitatively) that a substance causes adverse effects on an ecosystem as a result of environmental exposure. Both the magnitude, duration and frequency of the environmental exposure as well as the effects on inter- and intra specific relations of the target populations are ideally taken into consideration (EEC 1990a, Bro-Rasmussen 1988).

Risk Management (RM): Risk management of a chemical substance integrates the results of the hazard assessment or risk assessment with information regarding the costs and benefits of e.g.:

- the use of best-available technology for reducing the release of hazardous chemicals, substituting harmful chemicals with less harmful chemicals, improving waste water treatment

- other pollution control measures

Risk management may also include attempts to estimate the environmental impact of promising measures to reduce the concentration of the substance in the environmental compartment of concern. The measures to be taken are often considered case by case and include always economical and political desicions as well as scientific judgements. They might involve e.g. labelling, use or disposal instructions, emission control, pollution control requirements, restrictions on types of use, or even a total ban of the substance (cf. Table 12.2 in chapter 12.3).

12.2 Methods for the Assessment of the Bioaccumulation Potential of Chemicals

Within the different hazard and risk assessment schemes the assessment of the potential impact of a substance for the environment include the use of data regarding a number of intrinsic properties of the substance, i.e. data on ecotoxicity, environmental fate such as e.g. degradability and bioaccumulation.

For cost effective reasons, the estimation or determination of these intrinsic properties of chemicals are performed in successive steps in order to exclude the non concerning substances from comprehensive (and costly) investigations.

Following the recommendations given by OECD (1979), the cut off value for log P_{ow} triggering measurements of the BCF for fish are often set to log $P_{ow} = 3$, corresponding to BCF = 6-90 (cf. Figure 12.1, Esser and Moser 1982). For those substances with log $P_{ow} \geq 3$ bioconcentration studies on fish are in general recommended, especially if the substances are of concern due to high aquatic toxicity, persistency, high tonnage or special use as e.g. active ingredients in pesticides.

The term bioconcentration has in the past tended to be used synonymously with bioaccumulation in aquatic organisms as bioconcentration via direct uptake from water by gill breathing animals has shown to be the major route for various classes of xenobiotics.

The key-elements for the evaluation of the bioaccumulation potential of chemicals are:

- Determination of P_{ow}
- Correlations between P_{ow} and BCF

- Experimental determination of bioconcentration in fish
- Bioconcentration in fish as a predictor for bioconcentration in other aquatic species.
- Bioconcentration as a predictor for bioaccumulation/biomagnification in aquatic food chains.

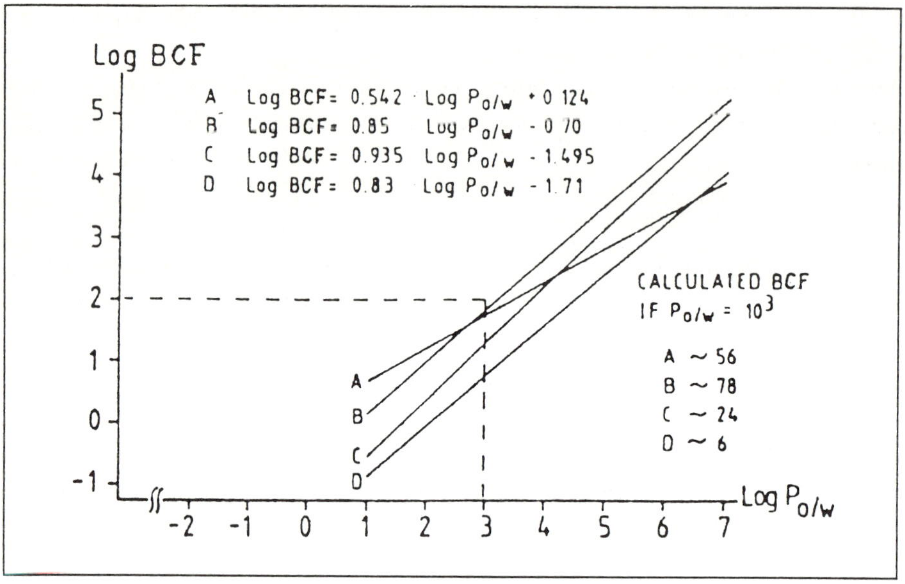

Figure 12.1 Log P_{ow} vs log BCF correlations for fish (slightly modified from Esser and Moser 1982).

It is outside the scope of this paper to discuss all items, which may influence the quality and predictability of various methods to assess the bioaccumulation potential of chemicals. Below are some of the major problems briefly discussed in relation to the use of the parameter "potential for bioaccumulation" in environmental hazard and risk assessment of chemicals in a regulatory context.

12.2.1 Estimation of the BCF by use of log P_{ow}

A number of reports have in the past shown linear correlation between log P_{ow} and log BCF for a variety of neutral organic substances.

These correlations have in general shown to be significant in the interval log P_{ow} 1 - 6 and have shown to increase in significans, when only data from substances of same chemical class are included. Up to date, however, only a limited number of chemical groups have been investigated for this correlation, e.g. chlorinated hydrocarbons and aromatic hydrocarbons (Connell 1990).

Although a number of the correlations have regression coefficients above $r = 0.9$, they may still contain pairs of data, which are far from being predicted by the regression.

Should log P_{ow}-log BCF correlations be applied for hazard assessment of chemicals, the risk of underestimating the potential of bioaccumulation must be at a low level. It is therefore most important to examine outliers e.g. to see whether they provide information about chemicals and their bioaccumulative kinetics for which the generalization is not valid.

More significant correlations is to be expected, when BCF values are based on fish lipid weight instead of whole body weight, which is most frequently reported. Normalizing BCF-values to the lipid content would reduce some of the variability caused by the inclusion of data from different species. "Real" outliers would thus be easier to identify.

Another point of concern is, that the confidence limit of the correlations is very seldom reported. This figure would allow regulators to apply a more "safe" estimate or range of BCF for a given P_{ow}-estimate.

As shown by e.g. Connell and Hawker (1988), Könemann and van Leeuwen (1980), Spacie and Hamelink (1982), a linear relationship between log P_{ow} and log BCF does not exist above approx. log P_{ow} 6. For superhydrophobic substances, experimental studies on bioaccumulation would be the only way for a "safe" prediction, although such substances are very difficult to handle experimentally due to their extremely low water solubility.

In summary log P_{ow}-log BCF correlations gives the order of magnitude of BCF-values for organic chemicals with a log P_{ow} between 1 and 6, when the bioconcentration is a simple partitioning phenomenon, i.e. for substances, which:

- are not (significantly) metabolized
- do not ionize in water
- do not have a molecular mass > approx. 700
- are not metals or organo-metals

12.2.2 Determination of the BCF

A number of test guidelines for the experimental determination of bioconcentration in fish have previously been adopted; the most widely applied being the OECD test guidelines and ASTM standard guide (OECD 1981, ASTM 1985).

In general, the principles of these guidelines are the same, i.e. to measure the bioconcentration in fish or bivalve molluscs (ASTM) during a period of exposure to the chemical dissolved in water. The potential of bioaccumulation is "approximated" by calculating the bioconcentration factor (BCF) as the concentration in whole fish (C_F) (or parts hereof) at "near-steady state" divided by the mean concentration of the chemical during the exposure period (C_W) and/or as the ratio between the rate constants of uptake (K_1) and depuration (K_2), assuming 1st order kinetics.

On the other hand, the experimental conditions of the individual guidelines are different, especially concerning:

- method of test water supply (static, semi-static or flow-through)
- the requirement for carrying out a depuration study
- the mathematical method for calculating BCF
- sampling frequency: no of measurements in water and no of samples of fish
- requirement for measuring lipid content of the fish
- the minimum duration of the uptake phase

The most widely applied assessment of chemicals is the OECD flow-through method (OECD Test Guideline no 305E). This method prescribes an uptake period followed by an elimination period with a duration twice that of the uptake period. Although an uptake period sufficient to reach steady state conditions is recommended, the uptake period (and thus the depuration period) may be designed based on log P_{ow} - K_2 correlations (K_2 = -0.414· log P_{ow} + 1.47, Spacie and Hamelink 1982) and thus calculating the expected time for e.g. 95 percent steady state ($3.0/K_2$) provided that the bioconcentration follows 1st order kinetics. C_F/C_W or the K_1/K_2-ratio may be applied for the calculation of BCF, although the former ratio is preferred.

The OECD Test Guideline 305E was ring tested in the EEC in 1985-1986 (Kristensen and Nyholm, 1987). Based on the experiences gained, the method was updated and since adopted by experts from the EEC member states. In 1988, the revised version was forwarded to the OECD Updating Panel for consideration of the replacement of the original OECD 305E-guideline.

Following circulation of this proposal, a number of questions were raised focusing especially on the following 3 areas:

- The validity of calculating the duration of the uptake period based on the 1st order kinetic model and thus the possibility of performing the uptake study in a shorter period than the 28 days required in the ASTM guide.

- The validity of estimating a steady-state bioconcentration factor based on K_1/K_2 calcu lations.

- The bioavailability of the test chemical in relation to the content of non-controlled particulate (and dissolved) organic matter in the test water.

Very few studies have been reported where other than 1st order kinetics have been observed or applied for calculating BCF.

For tetra-, penta- and hexachlorobenzene, the depuration from fish (Poecilia reticulata) showed clearly a biphasic pattern (Könemann and van Leeuwen 1980). The second slow part of the depuration phase did not show up until after 25 days of depuration. Further for tetra- and pentachlorobenzene this second phase was significantly observed only after approx. 50 days of depuration. For hexachlorobenzene an even longer depuration period would have been necessary for a reliable estimation of the two depuration constants. The results obtained by Könemann and van Leeuwen (1980) are shown in Figure 12.2.

In Table 12.1 the kinetic constants and BCF values (C_F/C_W) calculated on the basis of a two compartment model are compared with the BCFs expected if 1st order kinetics and the OECD test design are followed. The comparison indicates that despite the very clear 2 compartment kinetics for 2 of the 3 substances, the BCF-values obtained irrespective of the model applied are similar. Also the differences in BCF-estimates based on C_F/C_W and K_1/K_2 seem small. In a 28-day exposure period followed by a 14-day depuration period with 1,2,4,5-tetrachlorobenzene, a lipid-based BCF value of 5.010^{4} was estimated by 1st order K_1/K_2 calculations, Jordanella floridae being the test organism (Smith et al. 1990). This BCF-value is in excellent agreementwith the data obtained by Könemann and van Leeuwen (1980) for 1,2,3,5-tetrachlorobenzene.

As examplified by the study of Könemann and van Leeuwen (1980), the second of the two depuration phases will only be expected significantly to influence the K_2 (K_1)-estimate, when the relative amount of test chemical in the compartment showing slow depuration is high compared to the amount of chemical in the other compartment.

For substances, where a second compartment significantly influence the K_2 (K_1)-estimate, the slow second depuration phase would be expected to show up after a relative short time of depuration, and thus probably also to be noticed in tests following the OECD-update proposal.

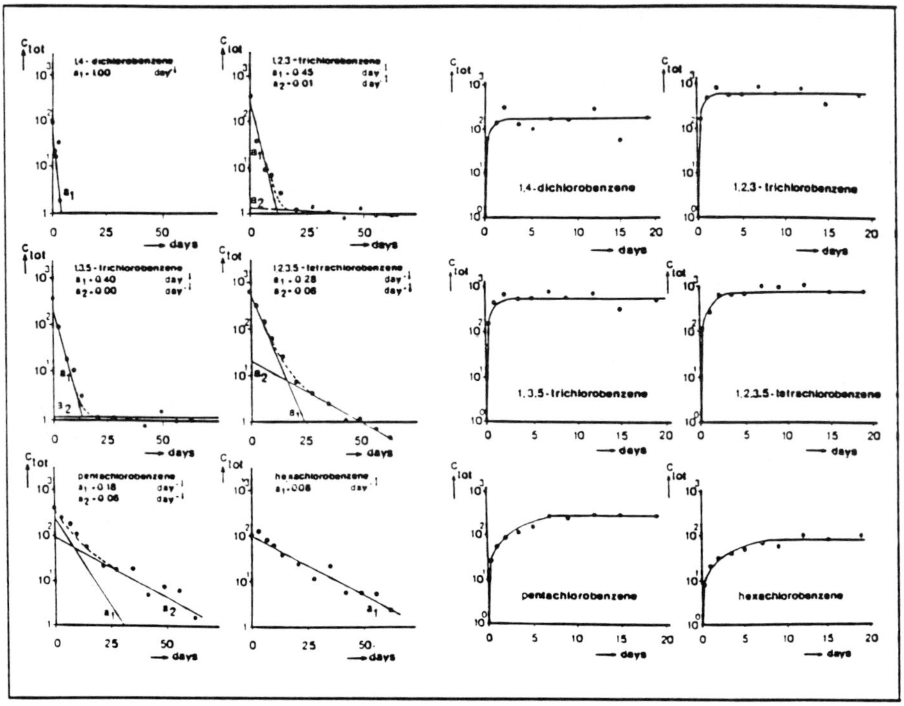

Figure 12.2 Residues of chlorobenzenes after accumulation and depuration. All concentrations (C_{tot}) are given as μg residue/g fat weight. (After Könemann and van Leeuwen 1980).

Applying 1st order kinetics, pentachlorobenzene and hexachlorobenzene would be expected to attain 95 pct of steady state equilibrium in a water-fish system after 23 and 50 days respectively (Table 12.1). Very often, however, an "apparent" steady state will be observed much earlier in uptake studies, because it is not possible to discriminate relatively small increases in concentration due to the variability between the sampled fish, errors in analytical procedures and the variability due to fluctuations of the concentration of the chemical in the test water. For e.g. pentachloro- and hexachlorobenzene "apparent" steady state after 7 days of exposure was observed (Könemann and van Leeuwen 1980). Thus, even if this study had been performed in accordance with the OECD-guideline no 305E regarding the time to reach 80 pct of steady state, the resulting BCF values would have been nearly identical.

The experience from the ring test of the OECD Test Guideline no 305E (1981) between 13

Table 12.1 Comparing data obtained by Könemann and van Leeuwen (1980) with data expected if the OECD proposal for an updated version of Test Guideline 305E (1988) had been applied.

SUBSTANCE	SOURCE	TEST DESIGN*				BCF***	
		log P_{OW}	K_2** (DAY^{-1})	EP (DAYS)	NEP (DAYS)	C_F/C_W	K_1/K_2
1,2,3,5-Tetrachlorobenzene	KÖNEMANN & VAN LEEUWEN	4.94	0.26	20	63	$7.2 \cdot 10^4$	$5.8 \cdot 10^4$
	OECD t_{80}		0.27	6	12	$\approx 7 \cdot 10^4$	$\geq 5 \cdot 10^4$
	OECD t_{95}			11	22	$\approx 7 \cdot 10^4$	$\geq 5 \cdot 10^4$
Pentachlorobenzene	KÖNEMANN & VAN LEEUWEN	5.69	0.11	20	63	$2.6 \cdot 10^5$	$2 \cdot 10^5$
	OECD t_{80}		0.13	12	24	$\approx 3 \cdot 10^5$	$\geq 1.2 \cdot 10^5$
	OECD t_{95}			23	46	$2.6 \cdot 10^5$	$\geq 1.2 \cdot 10^5$
Hexachlorobenzene	KÖNEMANN & VAN LEEUWEN	6.44	0.06	20	63	$2.9 \cdot 10^5$	$1.6 \cdot 10^5$
	OECD t_{80}		0.06	27	54	$2.9 \cdot 10^5$	$1.6 \cdot 10^5$
	OECD t_{95}			50	100	$2.9 \cdot 10^5$	$1.6 \cdot 10^5$

*: Test design, OECD update proposal: $\log K_2 = -0.414 \cdot \log P_{OW} + 1.47$
$t_{80} = 1.6/K_2$ (days), $t_{95} = 3.0/K_2$ (days), NEP (Non-exposure period) = 2 · EP (Exposure period)

**: OECD: K_2 calculated according to *), Könemann & van Leeuwen: K_2 calculated applying two compartment model.

***: C_F/C_W based on fat weight.

European laboratories (Kristensen and Nyholm 1987) showed relatively small deviations between K_1/K_2 and C_F/C_W estimates compared to the difference in BCF data derived from various laboratories.

The results from this ring test on Lindane are summarized in Figures 12.3 and 12.4 as estimated BCF values based on C_F/C_W and K_1/K_2 calculations (1st order kinetics, non linear regression). After having excluded experiments with highly fluctuating concentrations in the water and thus not in accordance with the quality criteria of the test guideline, the mean and standard deviation for 19 experiments was BCF = 450 ± 170 based on whole body dry weight and for 16 experiments, BCF = $1.1 \times 10^4 \pm 2.5 \times 10^2$ based on lipid weight.

For the purpose of an initial environmental hazard assessment of chemicals, therefore, an

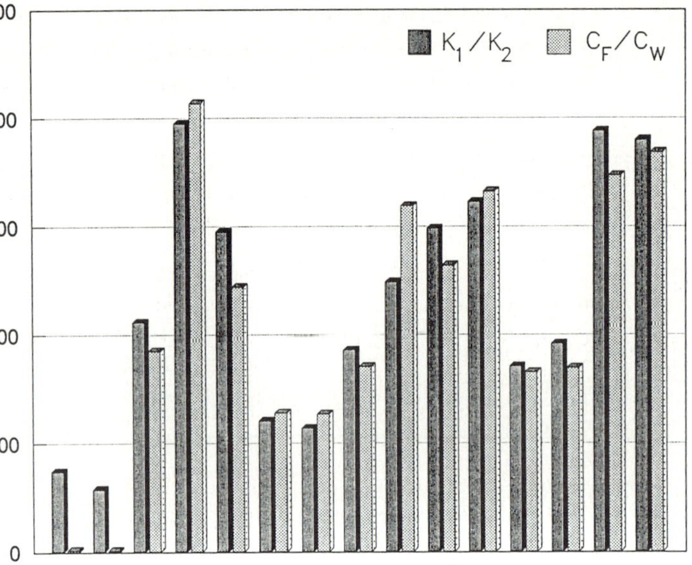

Figure 12.3 BCF - values for Lindane. Results of EEC ringtesting the OECD guideline 305 EE. C_F values based on whole body weight.

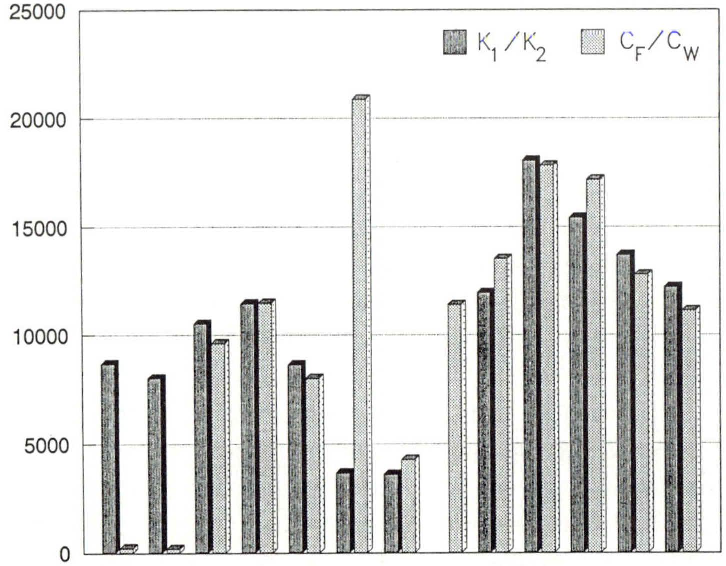

Figure 12.4 BCF - values for Lindane. Results of EEC rintesting the OECD guideline 305 E. C_F values based on lipid weight.

improved guideline (concerning the achievement of reproducibility) would have a greater effect on data quality and precision than would the selection of either K_1/K_2 or C_F/C_W, 1st or 2nd order kinetics, as preferred calculation methods. Another problem of concern is the bioavailability of the chemical in the test water. The standard approach in the varying test methods is to analyse samples of the test water for the total content of the test chemical at regular intervals during the uptake phase. It has often been reported (e.g. Servos et al. 1989, Opperhuizen and Stokkel 1988, Carlberg et al. 1986, Schrap and Opperhuizen 1990) that only the portion of chemical in "true" solution is bioavailable for uptake (and hence responsible for toxic effects). As fish are fed during bioconcentration studies, relatively high concentrations of dissolved (DOM) and particulate organic matter (POM) may reduce the fraction of chemical actually bioavailable for direct uptake via the gills. Applying the CF/CW-ratio, BCF may thus be underestimated, especially for lipofilic chemicals.

Recently, Scrap and Opperhuizen (1990) showed for HCB that BCF estimated as the K_1/K_2-ratio is in excellent agreement with the C_F/C_W-ratio, provided that CW is corrected for the amount of chemical sorbed to suspended organic material.

The difference between corrected and non-corrected BCF-values was for HCB a factor 5.8, and was calculated according to the following equation (Schrap and Opperhuizen 1990):

$$BCF = BCF^* \, y \, (1 + Kp \times S)$$

where

BCF:	bioconcentration factor based on "bioavailable" concentrations
BCF*:	bioconcentration factor based on actually measured total concentration in the test water.
Kp:	the sorption coefficient (l/kg)
S:	concentration of suspended organic material (kg dw/l)

For substances with log P_{ow} values below 4 - 5, the sorption to particulate matter did not influence the bioavailability significantly, as trichloro- and tetrachlorobenzenes resulted in nearly identical BCFs in experiments with and without high loadings of suspended organic material (deviation within a factor of 1.5) (Schrap and Opperhuizen 1990). For pentachlorobenzene (log P_{ow} 5.18), the BCF was reduced 2.4 times, when high loading of particulates was present. For superhydrophobic substances like hexa- and octachlorobiphenyls (log P_{ow} values of 6.90 and 7.11 respectively), suspended organic material did not influence the BCF in guppies. For tri- and tetrachlorobiphenyles (log P_{ow} = 5.5 - 5.8) however, significant reduction in BCF values were observed (particulate content 10 to 50 mg chromosorb per litre) (Opperhuizen and Stokkel 1988).

Also Carlberg et al. (1986) observed that trichlorobiphenyls were bioaccumulated less, when high concentrations of humic substances were present. It was further observed that the origin of the humic substances influenced the degree of this reduction.

The bioconcentration of naphthalene (log P_{ow} = 3.45) in bluegill sunfish was reduced 90% in the presence of dissolved humic acid (20 mg carbon/l), whereas benzo(a)pyrene (log P_{ow} = 6.50) was not affected (McCarthy and Jimenez 1985).

The lack of influence of particles on the bioconcentration factor in fish of superhydrophobic substances is assumed to be due to the very low solubility of these substances. Thus, dissolution rates, rather than the uptake rate into the organisms, are assumed to influence the bioconcentration process (Connell 1990).

Thus, both dissolved and particulate organic matter in the water may significantly influence the bioconcentration of organic chemicals in fish. This impact is especially influenced by the lipofilicity of the substance and the amount and type of organic matter.

For the calculation of the "steady state" BCF for fish, it is necessary to keep the concentration of the test chemical in the test water within relatively narrow limits according to the OECD and EEC Test Guidelines. It is thus necessary for certain lipofilic compounds also to focus on the content of particulate matter to keep the "**bioavailable**" concentration of the test chemical within relatively narrow limits.

BCF-data reported on lipofilic substances may in many cases have been underestimated when based on C_F/C_W-ratios from experiments with high content of particulate or dissolved organic matter. BCFs based on K_1/K_2-ratios are expected to be less affected by the degree of bioavailability, except for experiments where the content of organic matter has been highly fluctuating during the exposure period.

The problems regarding the importance of inter-laboratory variations, the way of BCF calculation, the use of 1st or 2nd order kinetics and the influence of DOM and POM have only been raised seriously in recent years, e.g. in relation to the proposal for an uptated version of the OECD Test Guideline no 305E.

Within the precision needed for initial environmental hazard assessment of chemicals, the OECD update proposal is expected to provide reliable BCF values for fish. The updated Guideline may in the future, however, be further improved especially regarding:

- Specific experimental measurements of the depuration constant (K_2) to be recommended, when 1st order kinetics are expected significantly to underestimate bioconcentration, i.e. prolonged depuration studies after e.g. **bolus** administration of the chemical

- General acceptance of BCF values based on fish lipid weight in parallel to fish total weight.

- Methods for correcting the influence by DOM and POM or methods for reproducible removal of DOM and POM from the test water.

12.2.3 Determination of the Potential for Bioaccumulation/Biomagnification

For comprehensive hazard and risk assessments, knowledge on the fate of chemicals are needed, which goes beyond bioconcentration studies in fish, i.e. the significans of contributions to bioaccumulation of chemicals through other compartments than water (e.g. via food, sediment). It is in general recognized, that substances with relative low log P_{ow}-values are not biomagnificated. For superhydrophobic substances (log $P_{ow} > 6$) bioaccumulation via food may however be significant (Connell 1990).

In general internationally recognized guidelines are lacking. At first, methods for measuring bioconcentration in other aquatic organisms than fish should be developed and agreed upon. Secondly, methods for determination of indirect bioaccumulation/biomagnification in bentic organisms is needed.

12.3 Use of Data on Bioaccumulation in formalized Environmental Hazard Assessment Procedures

The assessment of the bioaccumulation potential of chemical substances is normally performed as an element of an overall environmental hazard assessment or a risk assessment (cf. Table 12.2).

When an overall environmental hazard assessment of a substance is performed, the bioaccumulation potential is only one of a range of intrinsic properties of the chemical, which are used for the assessment. Also the environmental release of the substance is included. The most problematic stage(s) in the life cycle of the chemical are identified, i.e.:

- production/manufacturing;
- compounding (i.e. mixing with other chemicals making chemical preparations);
- industrial, agricultural or private use;
- recycling or disposal/incineration.

Table 12.2 Environmental hazard assessment. Modified from Bro-Rasmussen 1988.

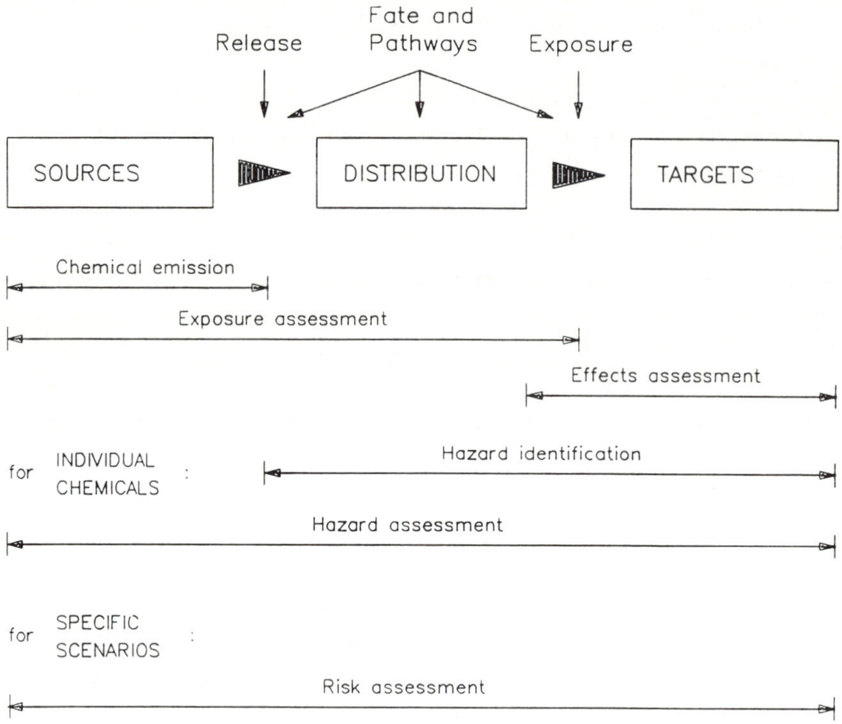

Here we will put most emphasis on how the potential for bioaccumulation of chemical substances are used together with other relevant intrinsic properties of chemicals for environmental hazard identification/assessment or environmental risk assesment, e.g. when considering the need of further ecotoxicological testing, further information on probable environmental release or before deciding to take regulatory measures. By doing this, we will describe how various, in some cases indicative, information on the potential for bioaccumulation of chemical substances, often are interpreted and used.

The importance of the bioaccumulation potential of chemicals has historically to a great extend been recognized, because persistent and bioaccumulating chemicals have exposed humans, as well as other animals at the end of the food web - such as birds and marine mammals - in some cases to an alarming or surprising extent. For this reason, the bioaccumulation potential of chemical substances might be interpreted as a parameter of both toxicological and ecotoxicological relevance. (As an example of the former it might be mentioned, that a special Risk-phrase "Danger of cummulation effects" is a part of the EEC health hazard classification system for labelling dangerous chemicals (EEC 1983, 1990c).

Nevertheless, we will here pay attention on the ecotoxicological relevance only, using selected examples of European environmental hazard assessment procedures.

In some of these, doubt is some times expressed, whether the potential for bioaccumulation is an effect- or an exposure/fate related parameter. This confusion is plausible, because bioaccumulation often is expressed through delayed toxic effects. These appear, when the concentration in the target organs of exposed organisms exceeds a certain no effect level after a lag phase, in which the build up of the substance takes place. Our interpretation is however, that the bioaccumulation potential of chemicals is a "pure" fate or exposure related parameter. The reason is, that bioaccumulation of a substance per se only reflects the concentration of the substance within biota.

12.3.1 EEC-Criteria for Classification and Labelling of Chemicals: "Dangerous for the Environment"

The EEC classification system consists of sets of criteria using physical/chemical properties, and toxicological and ecotoxicological data elements for chemical substances to be classified and labelled as dangerous within the meaning of the directive (dir. 79/831/EEC). For any dangerous substance danger symbol(s), and/or appropriate risk phrases and safety phrases are to be used on the label to advise the user of any relevant fire/explosion, health and/or environmental danger. Importers and producers of chemical substances have the responsibility to classify and label the substances according to the criteria stated.

For some chemicals the competent authorities of the EEC member states have agreed upon the appropriate health hazard classification and labelling, and the substances have been included within the list of dangerous substances (Annex I of the directive). At present a total of approximately 2000 substances are included in the list (EEC 1987). The list does not contain any substances classified for environmental danger, because the criteria for this have just recently officially been agreed upon. The criteria for classification and labelling chemical substances "Dangerous for the environment" have been included in a new "labelling guide", an Annex to the 6th Amendment of the Chemicals Directive (dir. 79/831/EEC). The criteria were formally approved by the EEC member states December 10, 1990 (EEC 1990c).

For preparations, consisting of mixtures of chemical substances, the classification criteria are prescribed in another directive (dir. 88/379/EEC (EEC 1988b)). Presently however, no criteria have been made regarding classification and labelling of preparations containing environmentally dangerous substances (Gustavsson and Ljung 1990).

Formalized classification criteria have only been agreed upon for chemicals dangerous for the aquatic environment. These consist of combination rules of simple parameters regarding the intrinsic properties of chemicals, i.e.

- the acute aquatic toxicity,
- the potential for biodegradation, and
- the potential for bioconcentration.

The lack of appropriate data for the vast majority of chemicals is the reason for not using more detailled data elements for classification. The data elements used within the classification criteria are regarded as the minimum appropriate for the environmental hazard identification. Are appropriate and more detailled data regarding the degradability in surface waters or the longterm toxicity towards aquatic organisms available, the classification scheme consists of certain "de-classification criteria" for some of the identified classes of substances (cf. below).

The formalizied criteria for classification and labelling chemical substances dangerous/ harmful for aquatic organisms are listed below:

- LC50 (fish, Daphnia, algae) $\leq$ 1 ppm
- LC50 (fish, Daphnia, algae) $\leq$ 10 ppm and not readily biodegradable and/or log P_{ow} ≥ 3
- LC50 (fish, Daphnia, algae) $\leq$ 100ppm and not readily biodegradable
- not readily biodegradable and log $P_{ow} \geq 3$

The minimum value of the short-term LC50-values regardless of the species is used. If a BCF-value in fish has been determined, this value has preference over the log P_{ow}-value. The cut off value is then BCF $\geq$ 100.

Substances classified according to criteria no 3 or 4 are not assigned the danger symbol "Dangerous for the environment" but only the appropriate risk phrase ("May cause long-term adverse effects for aquatic organisms"). Further these chemicals are not labelled if

- NOEC (fish or daphnia, long-term test) $\geq$ 1 ppm or
- it can be scientifically proved, that the substance is easily degradable in surface waters.

All of the data elements used for classification are to be determined according to the appropriate OECD Test Guidelines or with equivalent methods.

Also non-formal criteria are included within the set of classification criteria for the aquatic environment, thereby allowing classification on an ad hoc basis when appropriate.

Use of the log P_{ow}-value as an indicator for the bioaccumulative potential of organic substances especially for aquatic gill breathing animals is well established. Whether direct bioaccumulation from the water (bioconcentration) or food chain accumulation (biomagnification) in these animals dominate for superhydrophobic chemicals at realistic low environmental concentration is however debated (cf. e.g. Elgehausen et al. 1980, Spigarelli et al. 1982, Esser and Moser 1982, Moriarty and Walker 1987, Crossland et al. 1987, OECD 1987a, Connell 1990). In gill breathing animals the bioconcentration is normally regarded as a simple partitioning phenomenon, because the metabolic capacity of these animals are low compared to birds and especially mammals (Moriarty and Walker 1987, Walker 1990). Further the use of the log P_{ow}-value as an indicator of the potential for bioaccumulation in gill breathing animals is justified by empirically shown regression equations between the bioconcentration factor and the log P_{ow}-values for many classes of organic chemicals (cf. section 12.2).

Log P_{ow} = 3 has been agreed upon, as a cut off value for further testing of bioaccumulation in fish of organic chemicals (Esser and Moser 1982). The equivalent cut off value for BCF in fish has been set to 100 because a log P_{ow}-value of 3 approximates a BCFvalue of 6 - 90 in some of the reported regression equations of log P_{ow}-value vs. log BCF-values for organic chemicals (Figure 12.1).

The relative importance of the different parameters used within the EEC classification system are in decreasing order

- the acute toxicity towards aquatic organisms,
- the potential for biodegradation,
- the potential for bioaccumulation, and
- long-term aquatic toxicity/degradability in surface waters.

The inclusion of the data element "potential for bioaccumulation in aquatic gill breathing animals" (i.e. the log P_{ow}-value or the BCF-value for fish) within the set of criteria for classification and labelling of chemicals as harmful for aquatic organisms indicates, that the regulatory authorities in EEC recognize the need of a warning label on substances, which in addition to the bioaccumulation potential are of some ecotoxicological concern (i.e. being (acutely) toxic for aquatic organisms and/or not being readily biodegradable).

12.3.2 EEC Working Procedure for initial Hazard Assessment of new Chemicals

New chemical substances (not marketed within the EEC before september 18, 1981) are notified before being marketed according to "the 6th Amendment" (dir. 79/831/EEC), i.e. these chemicals are tested for various physical/chemical, toxicological and ecotoxicological properties (the "base set" of data is required) (EEC 1979 and 1984). Important "base set" information in this context are: the water solubility, the P_{ow}-value, the acute toxicity towards fish and Daphnia, the ready biodegradability of the substance and exposure related information regarding the intended marketed volumen, use and foreseen release pattern to the environment.

In certain concerning cases the Competent Authority (CA) from the member country, who receives a notification, may require immediate post base-set testing. When the marketed volumen of the test substance exceeds 10t/y per notifier (importer or manufacturer) the CA may require further testing according to "the level 1 tests" in Annex VIII of the directive, including prolonged fish and daphnia tests, further biodegradability testing, toxicity tests on plants and earth worms and experimental determination of BCF in fish (EEC 1988). Further testing has to be considered, when the marketed volumen exceeds 100 t/y per notifier. When the marketed level exceeds 100 or 1000 t/y per notifier, appropriate "level 2 tests" are conducted. These are also stated in Annex VIII of the Directive (EEC 1990a).

The overall purpose of environmental hazard assessment of the notified chemicals is to provide basis for requiring further appropriate information, including test data, or to take regulatory measures at an early stage of marketing (EEC 1990a).

A working procedure for making a common environmental hazard assessment in this context has recently been agreed upon by the EEC-member countries (a gentlemens agreement, not a formal decision, cf. EEC 1990a).

Our interpretation of this working procedure is shown in Table 12.3. At step one, the Environmental Hazard Identification stage of the procedure, notified chemicals of low concern are identified. Chemicals, which should be considered for further testing at the 100 t/y per notifier do not fulfill the criteria and are initially put aside (cf. Table 12.3, step 1).

One of the elements of the first step is to consider, whether the substance is classified "Dangerous for the environment". Notified chemicals of higher concern are evaluated at the next stages of the working procedure.

At step two of the working procedure an Initial Environmental Hazard Assessment is performed. If the predicted concentration of the substance "at the end of the discharge tube" to the aquatic environment (PIEC) exceeds 1 o/oo of the lowest L(E)C50value of the acute aquatic base set tests, a more comprehensive exposure assessment is performed using the "realistic worst case concept" with reference to the most important part of the life cycle of the chemical regarding the environmental release (cf Table 12.3, step 2).

Table 12.3 Principles of the EEC working procedure for the environmental assessment of new chemicals.

STEP 1. ENVIRONMENTAL HAZARD IDENTIFICATION:
Identification of new chemicals for initial hazard assessment.

---

1) **Criteria for classification:**

- health hazard classification :
  T_x, T, X_n R40, (45, 46, 47) or 48*
- Classified for environmental danger
  (cf. the text in section 3, criteria no 1 to 4)

(* : cf. EEC 1990c)

2) **Significant environmental release during manufactoring, use, recycling or disposal**

3) **Not inherently biodegradable** (cf. OECD 1987a, EEC 1988).

4) **"Suspected structure" (SAR)**

STEP 2. INITIAL ENVIRONMENTAL HAZARD ASSESSMENT:
PURPOSE/METHOD:

Principle:	Purpose:
$L(E)C_{50}$ min./PIEC < AF = 1000	comparison of presumed NEC(ecosystem) with estimated initial environmental conc.
$L(E)C_{50}$ min.: min. value of	
- fish (96 hrs) - algae (72 hrs) - Daphnia (48 hrs)	most sensitive species for acute tox. test on aquatic species to be used
AF: Application Factor	"translation" of $L(E)C_{50}$ min. to presumed NEC(ecosystem)
PIEC: Predicted Initial Environmental Concentration	generalized standard scenarios for point source release to aquatic ecosystem (river) for various use categories of chemicals

---

PIEC is used instead of PEC, because use of a general dilution factor for European surface waters is not reasonable (Greef & DeNijs 1990).

Use of "realistic local worst case assumptions" regarding no of release points, release factors and waste water volumen are used. For the reliable estimation of environmental release Industry Category/Use category Documents and information from industry are used. Only PIEC from point source release can be estimated.

Included within the initial environmental hazard assessment procedure is further an ad hoc evaluation of selected elements of "the base set", i.e. the shopping list of the items below are interpreted:

- the potential for bioaccumulation
- the slope of the "ecotoxicity curves"
- degradability
- SAR
- other toxic or mutagenic indications

How the potential for bioaccumulation could be evaluated is shown in Table 12.4.

For substances with log P_{ow}-values above 4, the equilibrium time for bioaccumulation of the substance in fish is increasing dramatically. In Table 12.4, a short hand method for choosing the appropriate (duration of a) prolonged fish toxicity test according to the log P_{ow}-value of the substance is presented. As shown by Hawker and Connell (1985 and 1988), the bioaccumulation factor of superhydrofobic substances (with log P_{ow}-values > 6) cannot be expected to follow the general regression equations between log P_{ow} and log BCF or log K_2. In an administrative context the BCF of these superhydrofobic substances should be determined experimentally in tests of justifiable duration, if necessary for the environmental hazard or risk assessment.

For lipofilic substances the next item on the "shopping list" should also be examined closely, i.e. the LC50-value for fish vs. time should be considered for presence of delayed toxic effects due to bioaccumulation of the chemical. In Table 12.5 the principles of such an evaluation is presented. Evaluation of the EC_{50}-values vs. time for Daphnia should be considered in the same way (Tyle 1990).

12.3.3 Danish Criteria for Approval of the Active Ingredient in Pesticides

In many countries comprehensive health and environmental assessments or risk assessments on pesticides are performed as a part of the risk management of these environmentally important chemicals, e.g. before approval of applications for the registration of the active ingredients in pesticides are granted by the authorities. More stringent requirements for registration of pesticides, were introduced in Denmark by the new Act on Chemical Substances and Products, which was approved by the Danish Parliament in 1987:

"Registration cannot be granted to substances or products which, in connection with their use or handling and storage are, or, on the basis of available studies or experience, are pre-

Table 12.4 Use of log P_{ow}-values for the estimation of BCF and for the selection of long-term fish toxicity test.

Log P_{ow}-value ≥ 3

a) Reconsider the log P_{ow} vs. log BCF regression equation and choose an equation of a class of substances to which the chemical belongs.
b) Do not extrapolate beyond the range of values measured to establish the regression equation.
c) Substances with:
- MW > 700 or
- calculated least diameter of the molecule > 5.5 Å or
- length of the molecule > 5.5 nm

are in generel not bioaccumulated although the log P_{ow}-value is greater than 3.
d) Consider whether the exposure period of the acute toxicity test on fish was of appropriate duration to allow for total bioaccumulation of the substance and subsequent immidiate manifestation of the lethal toxic effect, i.e. whether "the equilibrium LC_{50}-value" (e.g. t_{95} or $t_{95}+t_m$) might be significantly less than the experimentally determined LC_{50}-value (4 days) (cf. e.g. van Hoogen & Opperhuizen 1988, Hawker & Connell 1985, Connell 1990).

The duration of the uptake period until 95% of the total uptake:

Theoretical equation: $t_{95} = 3.0/K_2$.
Emperical equation: $\log K_2 = -0.414 \cdot \log P_{ow} + 0.122$

By combining these equations the time for reaching 95% of the equilibrium value of the concentration of the substance in fish, i.e. the t_{95}-period, can be calculated according to the log P_{ow}-value of the substance:

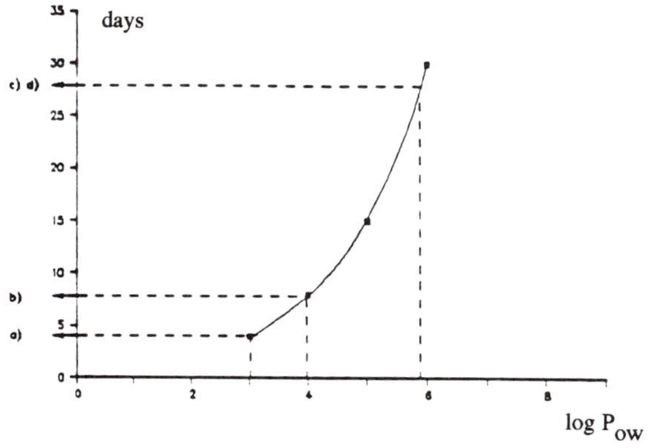

*: $t_{95} + t_m$, where t_m is the expression time for the manifestation of the intermediate toxic effect (here 2 days).
a) OECD Test Guideline No. 201 (Acute Fish Toxicity Test).
b) Danish/Dutch/Swedish proposal for an OECD Test Guideline, Short Term Fish Early Life Stage Test.
c) UK proposal for an OECD Test Guideline, Growth Inhibition Test on Juvenile Fish.
d) OECD Test Guideline, Fish Early Life Stage Test (1990).

Table 12.5 Evaluation of the slope of the acute ecotoxicity curves.

The L(E)C_{50}-time curves are interpreted:

The slope of the curves might indicate whether prolonged exposure of bioaccumulative chemicals might yield significantly lower L(E)C_{50}-values.

A flat time-effect curve in generel indicates, that this is not the case.

As a simple criteria for the evaluation, the ratio of the L(E)C_{50}-values within the observation period of the acute toxicity tests, could be used: e.g. for the fish toxicity test:

LC_{50} (72 hrs)/LC_{50} (96 hrs) > 1.25

and for the Daphnia toxicity test:

EC_{50} (24 hrs)/EC_{50} (48 hrs) > 2

could be used as a guidance. Cf. the figures below:

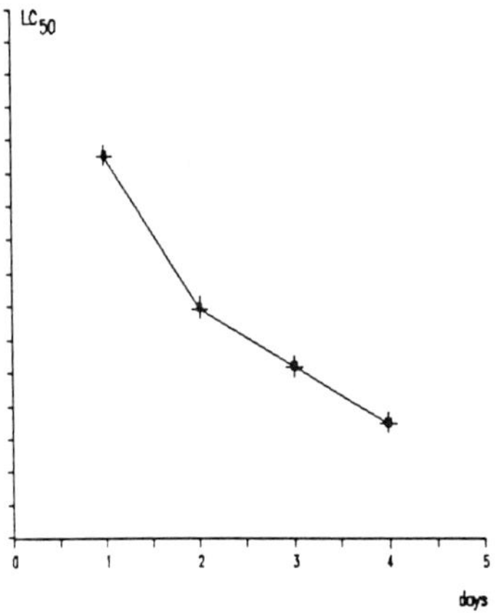

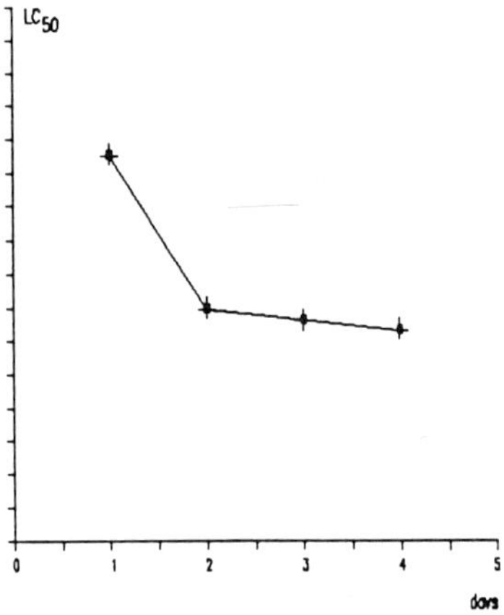

In **both** cases prolonged exposure **might** give rise to delayed effects in fish, if these show up after a certain lag time of longer duration than four days.

sumed to be especially dangerous to health or especially harmful to the environment".

The criteria described below were prepared by the National Agency of Environmental Protection (NAEP) in cooperation with the Pesticide Advisory Council and other experts.

The criteria are based on the assumption, that a pesticide shall be approved after an overall evaluation of the risk expected from its use to man and the ecosystem in which it spreads. Special attention is attached to certain health and environmental properties of the pesticides. It is, thus, taken into account that a property - either of the pesticide or its active ingredient - may alone or in combination with other properties be so undesirable, that the pesticide should not be used, even when protective measures are taken.

Criteria have been established for the properties considered to have a significant health or environmental impact, and for which the NAEP has had sufficient basic knowledge.

Criteria have been established for the following properties, which cause pesticides to be considered as especially dangerous to health or harmful to the environment:

1. Acute toxicity.
2. Toxicity by short term exposure.
3. Toxicity by prolonged exposure.
4. Carcinogenic effect.
5. Mutagenic effect.
6. Developmental toxicity.
7. Neurotoxicity.
8. Persistence in soil.
9. Mobility in soil.
10. Bioaccumulation.

Bioaccumulation is defined as "undesirable accumulation of active ingredients in living organisms in concentrations exceeding their concentration in the environment".

Below are specified the aspects relating to the active ingredient and its degradation products which are considered in the stepwise assessment procedure:

I) Evaluation of whether the substance is readily degradable in the environment.

A study is made to see whether the substance is "readily degradable in water" (within 28 days) according to the OECD tests of "ready biodegradability", or "readily degradable in soil" (half-life $\leq$ 20 days in the degradation test showing the longest degradation time) cf. the criteria for unacceptable persistence in soil (criteria 8).

If the substance is readily degradable it is not considered as a bioaccumulating substance. In case of doubt whether the substance is readily degradable or not, the evaluation at step II is performed.

II) Evaluation of the n-octanol/water partition coefficient (P_{ow}) indicating whether the substance has a potential for bioaccumulation.

If $P_{ow} < 1000$, the substance is not considered as bioaccumulating, unless other tests indicate the opposite (cf. steps III-V).

If $P_{ow} \geq 1000$, direct bioaccumulation shall be assessed (cf. step III). An assessment shall also be made to find indications of indirect bioaccumulation in mammals and/or other animals (cf. step IV).

III) Evaluation of the bioconcentration factor (BCF) for fish, expressing the potential for direct bioaccumulation.

This study is made if $P_{ow} \geq 1000$. BCF studies, if any, for other aquatic organisms shall be included in the evaluation. If contradictory P_{ow} and BCF values have been observed, the measured BCF shall prevail.

If BCF < 100, the substance is not considered as directly bioaccumulating.

IV) Indications of indirect bioaccumulation in mammals, birds and/or other animals shall be assessed on the basis of studies of the toxicokinetics (including metabolism) in rodents and/or residue studies in domestic animals or any other relevant study available.

Proper bioaccumulation tests shall be made if the estimated half-life of the substance is ≥ 3 days after one or more applications, calculated in relation to the compartment of the test animal, from which the substance depurates most slowly (cf. Trabalka et al. 1982, Trabalka and Gartens 1982).

A bioaccumulation study in mammals (step V) is required, even if the estimated $t_{1/2}$ in the toxicokinetic studies in mammals are borderline in relation to the criteria (i.e. apparent $t_{1/2}$ below, but not far below, 3 days.)

V) Studies of indirect bioaccumulation are usually made in mammals, and are, where required, designed in co-operation with the NAEP.

The dosing regime and the other conditions of the study should give results which allow a reliable estimation of the BCF. The study design is to be agreed upon by the NAEP.

An active ingredient is considered as indirectly bioaccumulating if x p.p.m. of the substance in the food results in more than x p.p.m. in a fatty tissue, i.e. if BF > 1.0. (BF = $C_{animal\ fat}/C_{food}$)

Bioaccumulation studies in birds/food chains are required, when significant exposure of these are probable and when other concerning evidence justifies this (cf. below)

A pesticide containing an active ingredient with a potential for bioaccumulation found in the studies described above cannot presently be registered in Denmark.

The cut off value of BF = 1 at step V of the evaluation procedure was chosen, because this value is a prerequisite for a proper biomagnification to occure. Further this cut off value seems to fit well with the cut off value for the log P_{ow}-value of 3 at step II and the cut off value of BCF(fish) = 100 at step III of the procedure (cf. Table 12.6 and 12.7), which are to be used to screen out those substances, which are not likely to bioaccumulate in mammals or birds (cf. Table 12.8, Trabalka et al. 1982, Trabalka and Garten 1982).

If a bioaccumulation test for the assessment of BF in birds is required, it might in most of the cases be appropriate to investigate the BF in a laboratory rodent species and then extrapolate the mammalian BF-value to the estimated level of BF in birds. (cf. Table 12.9)

The criteria for the assessment of the bioaccumulation potential have until now not been used for the denial of an application for the approval of any active ingredient in pesticides in Denmark.

For e.g. trifluralin, however, a use regulation has been required primarily because of the bioaccumulation potential of this substance in aquatic species, i.e. a 10 meter protective zone, where no spraying is allowed along surface waters, has been established. This warning is given in the instructions for use on the label of the pesticide. (According to the forwarded information on trifluralin the following BCF values for aquatic organisms have been observed, fish: 200 - 1600, Daphnia: approx. 600, algae: approx. 800. Further a field study on the bioconcentration in aquatic species and high toxicity towards aquatic organisms measured in long-term studies in the laboratory were regarded as supporting evidence.)

Table 12.6 Danish criteria for registration of pesticides: Potential for bioaccumulation. Assessment procedure.

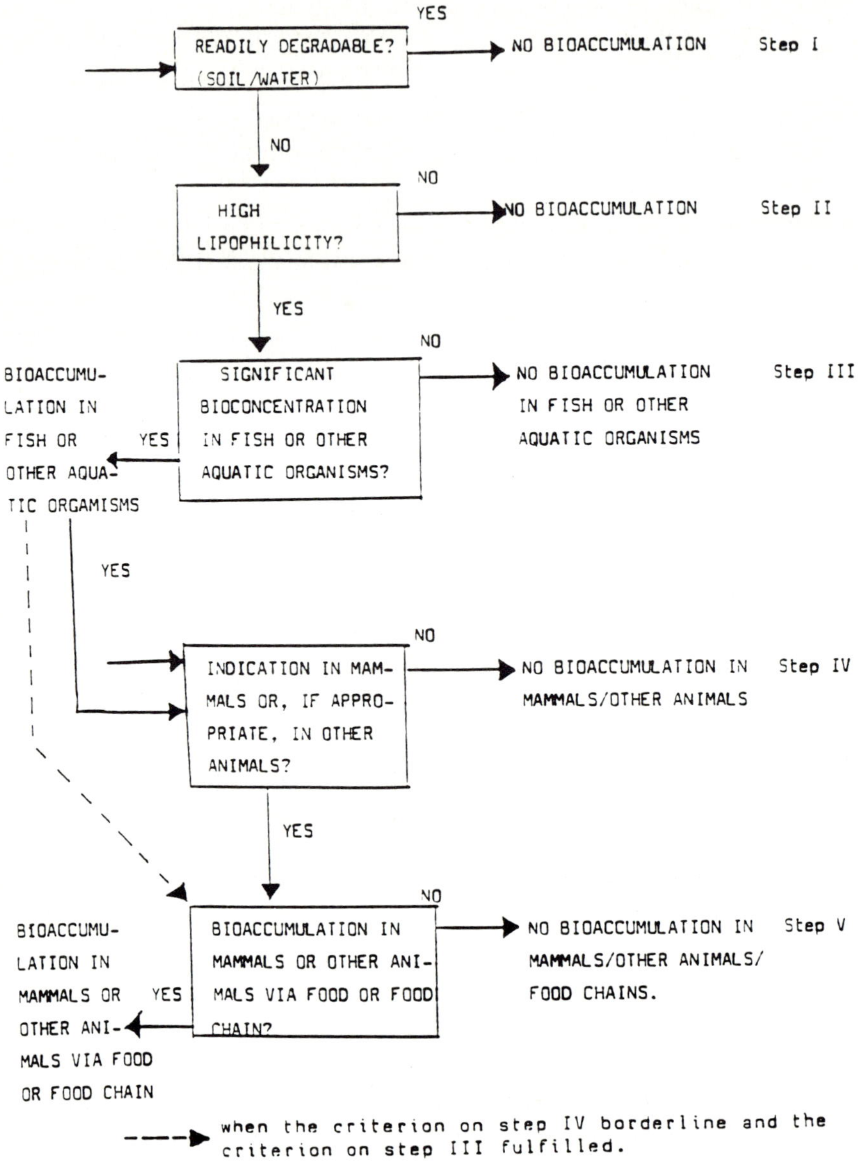

Table 12.7 Danish criteria for registration of pesticides: Potential for bioaccumulation. Assessment criteria.

I) "READILY DEGRADABLE":	"READILY DEGRADABLE" IN WATER ACCORDING TO ONE OF THE OECD TEST GUIDELINES OF READY BIO-DEGRADABILITY * OR READILY DEGRADABLE IN SOIL: $t_{1/2}$ IN SOIL $<$ 20 DAYS
II) "HIGH LIPOPHILICITY":	$P_{ow} \geqq 1000$
III) "SIGNIFICANT BIOCON-CENTRATION IN FISH":	BCF $\geqq$ 100 ACCORDING TO RESULTS FROM ONE OF THE OECD'S TEST GUIDELINES FOR DIRECT BIOACCUMULATION
IV) "INDICATION OF BIO-ACCUMULATION IN MAM-MALS ETC":	ESTIMATED $t_{1/2}$ IN THE COMPARTMENT OF THE ANIMAL WITH MOST INHERENT ELIMINATION $\geqq$ 3 DAYS
V) "BIOACCUMULATION IN MAMMALS ETC. VIA FOOD":	BF $>$1.0 IN TEST FOR INDIRECT BIOACCUMULATION

* c.f. the definition of ready biodegradability of the OECD "Guide-lines for Testing of Chemicals".

Table 12.8 Two-way cross tabulation for bioaccumulation factor in mammals and birds and the log P_{ow} based on paired observations for different chemicals.

Mammalian BFs: Two-way cross-tabulation for bioaccumulation factor (BF) in mammals and octanol-water partition coefficients (P_{ow}) based on 123 paired observations for 68 different chemicals.

Log BF (±0.5) \ Log P_{ow} (± 0.5)	-1	0	1	2	3	4	5	6	7
-5	1				1	6	1		
-4	1		1	1	5				
-3		2	6		8	5	3		
-2		2	2	6	6	3	1	1	
-1				1		1	3	3	1
0						3	10	18	2
1							6	9	2
2									1
3									

For mammals (ruminants and nonruminants) the overall regression equation eq. for the prediction of log BF(fat tissue) from log P_{ow} was:
log BF = $-3.972 + 0.583 \cdot \log P_{ow}$ ($P < 0.001$, $r^2 = 0.37$).

Bird BFs: Two-way cross-tabulation for bioaccumulation factor (BF) in birds and octanol-water partition coefficients (P_{ow}) based on 47 paired observations for 43 different chemicals.

Log BF (±0.5) \ Log P_{ow} (± 0.5)	-1	0	1	2	3	4	5	6	7
-4	1			1					
-3		2	1			2			
-2				1	3	2	1		
-1		1	2				2	1	
0						2		2	
1						1	7	10	3
2							1		1

For birds the overall regression equation for the predicted BF(fat tissue) was: log BF = $-2.743 + 0.542 \cdot \log P_{ow}$ ($P < 0.001$, $r^2 = 0.54$).
(from Trabalka et al. 1982).

Table 12.9 Correlations between log BF values in rodents and birds (poultry and small birds) and between rodents and mammals other than rodents (cattle, dogs, swine, sheep and primates) (from Trabalka et al. 1982).

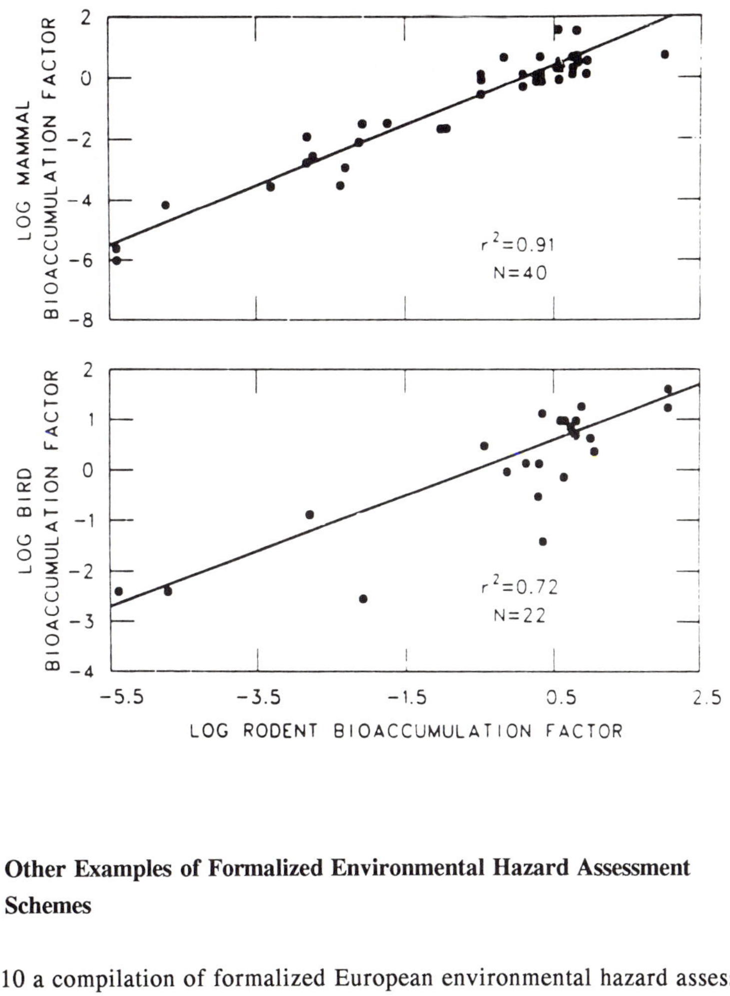

12.3.4 Other Examples of Formalized Environmental Hazard Assessment Schemes

In Table 12.10 a compilation of formalized European environmental hazard assessment schemes for various purposes is presented.

The potential for bioaccumulation is in the majority of schemes included by use of log P_{ow}- and BCF-values for fish and not by using a certain BF- or K_2-value for birds or mammals.

The reason for this is twofold. The (potential) bioconcentration in fish is regarded a significant parameter for the aquatic hazard assessment. Further, standard test methods for the determination of P_{ow} and BCF in fish, and thus P_{ow}- and BCF-data are available for a lot of chemicals (OECD 1984a, 1986, 1987a-d, 1988).

Table 12.10 Use of the parameter "potential for bioaccumulation" in selected formalized environmental hazard assessment systems.

EHA-SYSTEM	CNM	WMS	BUA	ESTHER
Methodology	Scoring system: low, medium,high + Combination rules for exposure, health and environmental effects	Scoring system: Individual scores for exposure rutes and effect types (10 combinations)	Scoring system on preselected existing chemicals of some environmental concern	Scoring system: 4 exposure indices (concern areas)
Purpose	Selection of existing chemicals for further assessment or testing			
Parameters	- production volume - P/C-properties - toxicity data - biodegradability - bioaccumulation - acute aquatic toxicity	environm. exposure: use, release, degradability, bioaccumulation etc. environm. effects: general toxicity aquatic toxicity	- occurence in the environment (W, S, T, a) - degradability (a, w) - bioaccumulation potential - acute mammalian toxicity - mutagenic/carciono-genic potential	Concern areas: - aquatic ecosystem - terrestial ecosystem - vertebrate predator - acute aquatic toxicity - wastewater treatment plant for each: use of relevant exposure and effect data
Trials of validation performed	Yes	?	Yes	Yes
Parameter for bioaccumulative potential	log P_{ow} (water solubility (S) and mw)	log P_{ow} or log BCF [mw > 300 log P_{ow} > 6] use BCF	log P_{ow} or log BCF	log P_{ow}
Cut off values	log P_{ow}: 3 & 5 (S < 25 ppm mw < 1000)	log P_{ow}: 2 & 4 (combined with data on pKa) log BCF: 1.5 & 3	log P_{ow}: 3 log BCF: approx. 2	log P_{ow}: 1 & 3 & 5
Relative importance of the bioaccumulative potential	For the concern area "environmental effects" the same as: - biodegradability - aquatic toxicity	Less than all other parameters	Same rel. importance as - acute aquatic tox. - acute mammalian tox. - mutagenicity/carcinogenicity potential	Only used for: - aquatic ecosystem - vertebrate predators relatively high scoring values
Use/status	?	?	In use	?
Origin	Nordic Countries	NL	FRG (Gesellschaft Deutscher Chemiker)	Sweden
Comments	Use of estimated log P_{ow} values discussed and validated	-	log P_{ow}-values calculated	-
References	Gjøs et al. 1989	Könemann & Visser 1983, 1988	GDCh 1989a, 1989b	Landner 1987 Gabring 1988 OECD 1984b

Table 12.10 continued

EHA-SYSTEM	EEC EHA	EHR/C&E
Methodology	EHI: ⌉ c.f. TABLE 3 & 4 EHA: ⌋ L(E)C50 / PIEC (river) compared with AP = 1000	Scoring system, exposures and effects dimension for the air, water and soil compartment
Purpose	Further testing/regulation of new notified chemicals (initial hazard assessment)	(hazard ranking)
Parameters	- exposure related data + P/C-data (PIEC-estimation) - acute aquatic toxicity - biodegradability - potential for bioaccumulation (base set of data)	Various data on: - toxicity (mammalian) - aquatic toxicity - mutagenicity - degradability - potential for bioaccumulation - exposure related data (base set of data + level 1 data)
Trials of validation performed	Yes	Yes
Parameter for bioaccumulative potential	log P_{ow} or log BCF	log P_{ow} or log BCF
Cut off values	EHI: log P_{ow} = 3; EHA: Guidance regarding duration of level 1 fish tests (BCF & toxicity) according to log P_{ow}	log P_{ow}: 1 & 2 & 3 log BCF: 1 & 2
Relative importance of the bioaccumulative potential	The least important of the intrinsic properties	air: 0 % water: approx. 15 % soil: approx. 10 % i.e. low
Use/status	In use as a working procedure in EEC	Used in France (working procedure)
Origin	CEC and EEC experts	FRG/France
Comments	Improvements on PIEC-estimation methods to be expected	Bioaccumulation both an exposure and effect parameter
References	EEC 1990a Tyle 1990	Jouany et al. 1982 Klein & Haberland 1982 Klein et al. 1986a & 1986b

Table 12.10 continued

EHA-SYSTEM	UK	NL	GESAMP
Methodology	4 choice trees (high, medium, low)	Several screening criteria (of equal importance)	Scoring or categorization; hazard profiles consisting of 5 different "columns"
Purpose	Selection of chemicals to be included within the "black list" (dir 76/464/EEC) and the "North Sea lists"		Control of marine pollution by bulk chemicals transported in ships
Parameters	a) acute aquatic toxicity b) chronic aquatic toxicity c) food chain bioaccumulation d) carcinogenicity	- marketed volume - carcinogenicity + persistence - long term aquatic toxicity + persistence - acute aquatic toxicity - persistence	- bioaccumulation/ tainting - damage to living resources - hazard to human health (oral) - hazard to human health (skin contact/inhalation) - reduction of amenities
Trials of validation performed	Yes	?	c.f. use/status
Parameter for bioaccumulative potential	log P_{ow} or log BCF	N.A.	Bioaccumulation to a significant extent (cf. below); bioaccumulating + risk to aquatic organisms/human; no data
Cut off values	?	N.A.	log P_{ow}: 3 $t_{1/2}$ (depuration) = 1 week
Relative importance of the bioaccumulative potential	Used in one of the choise trees	N.A.	One of 5 columns in the hazard profile
Use/status	?	?	List with hazard profiles on 2200 substances currently used
Origin	UK	NL	IMO/UN
Comments	-	Bioaccumulation might be included in an updated version	-
References	Jackson & Petersen 1990	Jackson & Petersen 1990	IMO 1987

The relative importance of the bioaccumulation potential in these hazard assessment systems is in most cases less than the relative importance of data elements regarding toxicity, degradation and environmental release/exposure.

Toxicity is unlike the potential for bioacumulation directly an indication of the possibility of adverse environmental effects. Further significant exposure and high persistence are prerequisites for bioconcentration to occure. Finally log P_{ow}- and BCF values can only be used for the estimation of bioconcentration in gill breathing animals and as a screening parameter to sort out those substances, which are not likely to bioaccumulate in food chains.

Therefore the relatively low significance of the parameter "potential for bioconcentration" (i.e. the log P_{ow} or BCF value) in the formalized environmental hazard assessment systems seems justified.

12.4 Conclusions

The future direction of standard test method development for environmental fate and effects of chemicals and environmental hazard assessment strategies for various administrative purposes are at the moment subject to detailed discussions within EEC and OECD working groups.

For assessment of the bioaccumulation potential of chemicals, test guidelines on P_{ow}-measurements and fish bioconcentration test are available. A new and significantly improved guideline for the latter has recently been proposed to the OECD Test Guideline programme. urgently needed.

Information on bioconcentration of chemicals in fish can only be used to sort out those chemicals, which are not likely to biomagnificate in higher vertebrates and to indicate warnings on potentials for high exposure concentrations in the biota.

Test guidelines on indirect bioaccumulation (biomagnification) are needed, because prediction of food chain accumulation of chemicals is considered important in formalized hazard assessment schemes and for ad hoc assessments of the environmental hazards of certain chemicals such as pesticides and high priority pollutants.

Available evidence suggests that information on indirect bioaccumulation of chemicals in rodents could be used as predictor of bioaccumulation in birds and mammals. Therefore it should be considered whether the test guidelines for long term (i.e. subchronic/chronic) toxicity studies in the laboratory rat could be supplemented with a section on how to measure the bioaccumulation of suspected test chemicals.

12.5 References

ASTM. American Society for Testing and Materials 1985. Standard Practice for Conducting Bioconcentration tests with fishes and Saltwater Bivalve Molluscs. E 1022-84.

Bro-Rasmussen, F. (1986). Application of Protocols for Hazard Assessments. Lecture held at the International Congress of Attitudes to Toxicology in the European Economic Communities, sponsored by European Society of Toxicology. EST, Tenerife 5.- 8. December 1986.

Bro-Rasmussen, F. (1988). Ecological Effects Assessment - An Overview. Prepared for the OECD Workshop on Ecological Effects Asessments. Washington D.C. 13. - 17. June 1988.

Carlberg, G.E., Martinsen, K., Kringstad, A., Gjessing, E., Grande, M., Källqvist, T. and J.U. Skaare (1986). Influence of aquatic humus on the bioavailability of chlorinated micropollutants in Atlantic Salmon. *Arch. Environ. Contam. Toxicol.*, 15, 543-548.

Connell, D.W. and D.W. Hawker (1988). Use of Polynominal Expressions to Describe the Bioconcentration of Hydrophobic Chemicals in Fish. *Ecotoxicol. Environ. Safety*, 16, 242-257.

Connell, D.W. (1988). Bioaccumulation Behaviour of Persistent Organic Chemicals with Aquatic Organisms. *Rev. Environ. Contam. Toxicol.*, 101, 131-148.

Connell, D.W. (1990). Bioaccumulation of xenobiotic compounds CRC Press. Inc., Boca Raton Florida, USA. pp. 219.

Crossland, N.O., Bennett, D. and C.J.M. Wolff (1987). Fate of 2,5,4'-Trichlorobiphenyl in Outdoor Ponds and Its Uptake via the Food Chain Compared with Direct Uptake via Gills in Grass Carp and Rainbow Trout. *Ecotoxicol. Environ. Safety*, 13, 225-238.

EEC (1979). Council of the European Communities. Council Directive of 18 September 1979 amending for the sixth time Directive 67/ 548/EEC on the approximation of the laws, regulations and administrative provisions relating to the classification, packaging and labelling of dangerous substances. dir. 79/831/EEC.

EEC (1983). Commission of the European Communities. Classification and Labelling of Dangerous Substances. dir. 83/467/EEC.

EEC (1984). Commissions Directive of 25 april 1984 adapting to the technical progress of the 6th Council Directive 67/548/EEC on the approximation of the laws relating to the classification, packaging and labelling og dangerous substances. Official Journal of the European Communities L 251, 27 (dir. 84/449/EEC, Base Set Test Methods of dir. 79/831/EEC).

EEC (1986). EEC Council Directive of 4 May 1986 on pollution caused by certain dangerous substances discharged into the aquatic environment of the community (dir. 76/464/EEC).

EEC (1987). Commission of the European Communities. Legislation on Dangerous substances. Classification and labelling in the European Communities. Consolidated text of Council Directive 67/548/EEC. Vol. 1 and 2.

EEC (1988). Commission directive of 18. november 1987 adapting to technical progress of the 9th time Council Directive 67 548 EEC on the approximation of laws, regulations and administrative provisions to the classification, packaging and labelling of dangerous substances. Official Journal of the European Communities L 133, 31 (dir. 88/302/EEC; i.e. Level I Test Methods according to Annex VIII, dir. 79/831/EEC).

EEC (1990a). Hazard and Risk Assessment in the Context of Directive 79/831/EEC. Workshop on Environmental Hazard and Risk Assessment in the Context of Directive 79/831/EEC. Working Doc XI/730/rev.3 (Presented by P.Greier) ISPRA, 15-16 october.

EEC (1990b). Directive 76/464 and North Sea Declaration. Seminar on "The Setting of a Common Selection Scheme of Dangerous Substances". Final Report.

EEC (1990c). Doc. XI/410/90. Draft Commission Directive of 10.12.1990 adapting to technical progress for the twelfth time Council Directive 67/548/EEC on the approximation of the laws, regulations and administrative provisions relating to the classification, packaging and labelling of dangerous substances.

Ellgehausen, H., Gruth, J.A. and H.O. Esser (1980). Factors determining the bioaccumulation potential of pesticides in the individual compartments of aquatic food chains. *Ecotoxicol. Environ. Safety*, 4, 134-157.

Esser, H.O. and P. Moser (1982). An Appraisal of Problems Related to the Measurement and Evaluation of Bioaccumulation. *Ecotoxicol. Environm. Safety*, 6, 131-148.

Gabring (1988). Preliminary assessment of the environmental hazard of chemical substances. An evaluation of the "ESTHER"-manual. Kemi Rapport från Kemikalieinspektionen. Utredningsavdelningen (in Sweedish).

GDCh-Advisory Committee on Existing Chemicals of Environmental Relevance 1989a. Existing Chemicals of Environmental Relevance. Criteria and List og Chemicals. Beratergremium für Umweltrelevante Altstoffe (BUA).

GDCh-Advisory Committee on Existing Chemicals of Environmental Relevance 1989b. Existing Chemicals of Environmental Relevance II. Selection Criteria and Second Priority List. Beratergremium für Umweltrelevante Altstoffe (BUA).

Gjs, N., Mller, M., Hægh, G.S. and K. Kolset (1989). Existing chemicals: Systematic data collection and handling for priority setting. Miljrapport 1989:8, Nordisk Ministerråd pp. 53.

Gobas, F.A.P.C. and D. Mackay (1987). Dynamics of Hydrophobic Organic Chemical Bioconcentration in Fish. *Environ. Toxicol. Chem.*, 6, 495-504.

Greef, J. and A.C.M. DeNijs (1990). Risk Assessment of New Chemicals. Dilution of Effluents in the Netherlands. Report No. 670208001. RIVM. Bilthoven (May 1990).

Gustafsson, L. and E. Ljung (1990). Substances and Preparations Dangerous for the Environment. A System for Classification, Labelling and Safety Data Sheets. Final report from a Nordic Working Group. Nordic Council of Ministers, June, 66 pp.

Hart, J.W. and N.J. Jensen (1990). The Myth of the Final Hazard Assessment. *Regulatory Toxicology and Pharmacology*, 11, 123-131.

Hawker, D.W. and D.W. Connell (1985). Relationships between Partition Coefficient, Uptake Rate Constant, Clearance Rate Constant and Time, to Equilibrium for Bioaccumulation. *Chemosphere* ,14(9), 1205-1219.

Hawker, D.W. and D.W. Connell (1988). Influence of Partition Coefficient of Lipophilic Compounds on Bioconcentration Kinetics with Fish. *Water Res.*, 22(6), 701-707.

IMO (1987). Composite list of hazard profiles of substances carried by ships. Joint Group of Experts on the Scientific Aspects of Marine Pollution (GESAMP), T4/7.01, 22 January.

ISPRA (1990). Summaries of Presentations. Workshop on Environmental Hazard and Risk Assessment in the Context of Directive 79/ 831/EEC.

Jackson, J. and P.J. Petersen (M.A.R.C., Campden Hill, London W8 7AD) (1989). Evaluation of Selection Schemes for Identifying Priority Aquatic Pollutants. Final Report - Study contract No B 6612/290/89 for the EEC Commission/DG XI.

Jouany, J.M. et al. (1982). Approach to Hazard Assessment by a qualitative system based on interaction concept between variables, pp. 367-86. Chemicals in the Environment. Chemicals Testing and Hazard Ranking - The Interaction between science and administration. International Symposium Lyngby-Copenhagen-Denmark 18.10.1982-20.10.1982. Proceedings. Eds. K. Christiansen, B. Koch and F. Bro-Rasmussen.

Klein, A.W. and W. Haberland (1982). Environmental Hazard Ranking of New Chemicals Based on European Directive 79/831/EEC, annex VII, pp. 419-34. Chemicals in the Environment. Chemicals Testing and Hazard Ranking - The Interaction between science and administration. International Symposium Lyngby-Copenhagen-Denmark 18.10.1982-20.10.1982. Proceedings. Eds. K. Christiansen, B. Koch and F. Bro-Rasmussen.

Klein, A.W. et al. (1986a). Weiterentwicklung eines Verfahrens zur Einstufung neuer Stoffe hinsichtlich ihrer Umweltgefärlichkeit. Teilbericht 1. Umweltforschungsplan des Bundesministers für Umwelt, Naturschutz und Reaktorsicherheit. Forschungsbericht 106 04 022. Oktober 1986.

Klein, A.W. et al. (1986b). Weiterentwicklung eines Verfahrens zur Einstufung neuer Stoffe hinsichtlich ihrer Umweltgefärlichkeit. Teil II: Gesamtdarstellung in englischer Sprache. Umweltforschungsplan des Bundesministers für Umwelt, Naturschutz und Reaktorsicherheit. Forschungsbericht 106 04 022. Oktober 1986.

Kristensen, P. and N. Nyholm (1987). Ringtest Programme 1984-1985: Bioaccumulation of Chemical Substances in Fish, Flow Through Method. Commission of the European Community, Degradation/Accumulation Subgroup. March 1987. (Final Report).

Könemann. H. and K. van Leeuwen (1980). Toxicokinetics in fish: Accumulation and elimination of six chlorobenzenes by Guppies. *Chemosphere*, 9, 3-19.

Könemann, H. and R. Visser (1983). Netherlands approach for setting environmental priorities for giving attention to existing chemicals: WMS-scoring system. Working document pp. 47.

Könemann, H. and R. Visser (1988). Selection of chemicals with high hazard otential: Part 1:WMS-Scoring System. *Chemosphere*, 17 (10), 1905-1919.

Landner, L. (1987). Kemiske ämnens miljöfarlighet. Manual för inledande bedömning. "ESTHER". Slutrapport nr. 1 från projektområdet ESTHER.

Lundgren, A. (1989). Comparison of different models for Environmental Hazard Classification of Chemicals. A status Report from the Joint Nordic Project "Guidelines for Environmental Hazard Classification of Chemicals". Kemikalieinspektionens rapport No. 9/1989.

McCarthy, J.F. and B.D. Jimenez (1985). Reduction in bioabailability to bluegills of polycyclic aromatic hydrocarbons to dissolved humic material. *Environ. Toxicol. Chem.*, 4, 511-521.

Ministry of the Environment, Denmark. National Agency of Environmental Protection, April 1987, October 1988. Criteria for Registration of Pesticides as especially dangerous to health or especially harmful to the environment.

Moriarity, F. and C.H. Walker (1987). Bioaccumulation in Food Chains - A Rational Approach. *Ecotoxicol. Environ. Safety*, 13, 208-215.

OECD (1979). Chemicals Testing Programme Expert Group. Degradation/Accumulation. Umweltbundesamt, Bundesrepublik Deutschland and Government of Japan. Berlin and Tokyo. December 1979. (Final Report).

OECD (1984a). DIGs Data Interpretation Guides for Initial Hazard Assessment of Chemicals. Provisional. DIG 5, 6 and 7. Paris.

OECD (1984b). Chemicals Group and Management Committee. Selected Approaches to Data Integration in Assessments of Chemicals. ENV/CHEM/CM/84.14. Environmental Directorate, OECD, p. 61 (prepared by L. Landner).

OECD (1986). Existing chemicals. Systematic Investigation. Priority Setting and Chemicals Review. Paris.

OECD Guidelines for Testing of Chemicals (1987a). Organisation for Economic Co-operation and Development. Summary of Considerations in the Report from the OECD Expert Group on Degradation and Accumulation.

OECD Guidelines for Testing of Chemicals (1987b). Organisation for Economic Co-operation and Development. Partition Coefficient (noctanol/water), High Performance Liquid Chromatography (HPLV) Method. Adopted 30.03.89, No 117.

OECD Guidelines for Testing of Chemicals (1987c). Organisation for Economic Co-operation and Development. Partition Coefficient (noctanol/water). Adopted 12.05.81, No. 107.

OECD Guidelines for Testing of Chemicals (1987d). Organisation for Economic Co-operation and Development. Bioaccumulation: Flow through fish test. Adopted 12.05.81, No. 305E.

OECD (1988). Updating proposal for a revision of Guideline 305 E: Accumulation: Flow through test. September 1988.

OECD (1990). Environment Monographs No. 35. A Survey of New Chemicals Notification Procedures in OECD Member Countries, June 1990.

Opperhuizen A. and R.C.A.M. Stokkel (1988). Influence of contaminated particles on the bioaccumulation of hydrophobic pollution. 51:165-177.

Schrap, S.M. and A. Opperhuizen (1990). Relationship between bioavailability and hydrophobicity: Reduction of the uptake of organic chemicals by fish due to the sorption on particles. *Environ. Toxicol. Chem.*, 9, 715-724.

Servos, M.R., Muir, D.C.G. and G.R. Barrie Webster (1989). The effect of dissolved organic matter on the bioavailability of polychlorinated dibenzo-p-dioxins. *Aquatic Toxicol.*, 14, 169-184.

Slooff, W., van Oers, J.A.M. and D. de Zwart (1985). Margins of Uncertainty in Ecotoxicological Hazard Assessment. *Environ. Toxicol. Chem.*, 5, 841-852.

Smith, A.D., Bharath, A., Hallard, C., Orr, D., McCarty, L.S. and G.W. Ozburn (1990). Bioconcentration kinetics of some chlorinated benzenes and chlorinated phenols in American Flagfish, Jordanella floridae. *Chemosphere*, 20(3/4), 379-386.

Spacie, A. and J.L Hamelink (1982). Alternative Models for Describing the Bioconcentration of Organics in Fish. *Environ. Toxicol. Chem.*, 1, 309-320.

Spigarelli, S.A., Thommes, M.M. and A.L. Jensen (1982). Prediction of Chemical Accumulation by Fish. EPA PB8 4 156918. January 1982.

Trabalka, J.R. et al. (1982). Xenobiotic Bioaccumulation by Terrestrial Vertebrates: A bibliography for food-chain modeling. Contract No. W-7405-eng-26. Environmental Sciences Division, Publication No. 1855. ORNL-5829, DE 82 015854.

Trabalka, J.R. and C.T. Garten, Jr. (1982). Development of Predictive Models for Xenobiotic Bioaccumulation in Terrestrial Ecosystems. Contract No. W-7405-eng-26. Environmental Sciences Division, Publication No. 2037. ORNL-5869, DE 83 003171.

Tyle, H. (1990). Base Set Information to be Considered beyond the Elements used for Initial Hazard Identification and Assessment. Presented at the meeting on Hazard Assessment at ISPRA, Italy; 15-16 October, 1990. Working document.

van Hoogen, G. and A. Opperhuizen (1988). Toxicokinetics of Chlorobenzenes in Fish. *Environ. Toxicol. Chem.*, 7, 213-219.

Walker, C.H. (1990). Persistent Pollutants in Fish-eating Sea Birds - Bioaccumulation, Metabolism and Effects. *Aquatic Toxicol.*, 17, 293-324.

Final Considerations

Roland Nagel

13.1 Introduction

At the beginning of this meeting, Böhling and Loskill arranged the purpose of the workshop in three areas (validation, extrapolation and evaluation) and characterized them by various questions. Now I am going to try to summarize and give some answers, and this will include, as far as possible, the results of the discussion.

13.2 Validation

Do the present test guidelines yield reliable results?

Besides the test guidelines of the EPA, the OECD test guidelines (Table 4.1) are of particular importance. In practice, the guidelines 305 C and 305 E are most frequently used (Caspers and Schüürmann 1991; Kristensen and Tyle 1991). The OECD aims to work out a comprehensive revision of their guidelines, especially with regard to the regulation of the mode of exposure. It has to be considered that in most cases the more precise and reproducible results are achieved under flow-through conditions. In general one runs the risk of measuring lowered values of bioconcentration factor when the assumed water concentration of the chemical does not correspond to bioavailability. This holds especially true for lipophilic compounds when test concentration exceeds water solubility. A further problem is given by the fact that the comparability of results originating from different laboratories can not be assumed (Kristensen and Tyle 1991).

For very lipophilic substances it might occur that the steady state cannot be reached within the test period. Applying the kinetic method for bioconcentration factor determination means that the choice of the mathematical models is particularily important in this situation.

How can the modeling of bioconcentration be performed?

The mathematical description of bioconcentration is mainly based on compartment models, as reviewed by Spacie and Hamelink (1982). The use of two- or multi-compartment models has consequences for the evaluation. The bioconcentration factor will not change drastically by using different models (Kristensen and Tyle 1991). This holds true at least for the situation when the elimination period is started at steady state conditions (Butte 1991). However, differences arise when the residue of a chemical in an organism is being assessed (Nagel 1988, Butte 1991). In the case of a very slow elimination from a large second compartment for instance, the assessment of the retention using the one-compartment model will be wrong.

How accurate are log P_{OW}-log BCF correlations?

The approach of QSARs, estimating the BCF from water solubility or from the octanol-water partition coefficient, is based on presumptions of limited validity. Firstly, a distribution and equilibrium model is assumed, which is a very simplified description of the biological processes involved. Secondly it is assumed that octanol is a suitable surrogate for the lipid phase of organisms (for review see Hawker and Connell 1989, compare Chiou 1985), which is not generally accepted (Dobbs and Williams 1983, Opperhuizen et al. 1988, Hawker 1990). Opperhuizen (1991) suggested to use the activity coefficients in water as correlation parameters.
The variability of log BCF-log P correlations and the possibility of large errors using QSARs show, that the estimation of BCFs for assessment purpose based on partition coefficients is questionable. The "worst case curve" suggested by Nendza (1991) has advantages over linear regression functions normally used. Caspers and Schüürmann (1991) proposed QSARs for distinct classes of compounds.

13.3 Extrapolation

Can laboratory results be extrapolated to environmental conditions and vice versa?

Under environmental conditions, many parameters are acting on bioaccumulation to different extent and without being quantitatively valuable. The bioconcentration factor measured in the laboratory is therefore not transferable to the situation in the field, and the residues measured in field studies may be much more higher.

Is a distinction between bioconcentration and biomagnification necessary?

For lipophilic compounds the elevated residues measured in the field compared with data from the laboratory can be explained by the accumulation via food. Stephan (1985) suggested to establish a bioaccumulation factor in order to evaluate total bioaccumulation. This bioaccumulation factor includes the accumulation of the fraction of the chemical which is absorbed on food particle, after feeding until steady state achievement. However, the biomagnification of less lipophilic compounds can be judged as being negligible (Opperhuizen 1991).

Can test results be extrapolated between organism groups?

Special problems arise concerning the interspecies extrapolation. With regard to non-degradable, persistent lipophilic chemicals, serving as a model for the ideal case, the transferability is well founded (Connell 1991). In practice, however, enormous differences have been found, even regarding lipid based values. Moreover in species-specific accumulation, unexpectedly high residue values may be obtained. For example, Hryk (1990) found in case of phenol accumulation by a water snail (*Lymnea stagnalis*) biphasic uptake kinetics, and 28 days of exposure resulted in 195fold enrichment of phenol without steady state achievement. Wänke (1991) determined a BCF of 354 for the bioconcentration of phenol in *Daphnia magna*. These values exceed fish bioconcentration factors (about 10) several times. As a consequence, supplementary information on bioaccumulation is required. Kristensen and Tyle (1991) request additional guidelines, for example a test with a benthic organism.

13.4 Evaluation

Does bioaccumulation reflect toxicity?

According to McKim and Schmieder (1991) bioaccumulation is not reflective of toxicity. It merely informs about the activity of the chemical and how it will act kinetically. Whether or not a chemical is highly bioaccumulatable says nothing about its toxic potency.

How can data on bioaccumulation be evaluated differentially, which current approaches to assess bioaccumulation exist and are they considered adequate to undergo further development?

An essential part in environmental hazard assessment of chemicals is the comparison between exposure and effect data. Kristensen and Tyle (1991) arranged bioaccumulation as a parameter of exposure in a scheme of environmental hazard assessment. However, the relevance of bioaccumulation of chemicals in organisms, independent from possible effects, justify an isolated assessment (for example Mearns 1985). Compared with other parameters, for example toxitiy, bioaccumulation is less important (Kristensen and Tyle 1991). Several formalized assessment schemes have been developed which combine the relevant parameters

(Frische et al. 1979, Korte 1980, Halfon and Reggiani 1986, Maki 1986, overview in Kristensen and Tyle 1991). In the meantime the Federal Environmental Agency has also presented an assessment scheme (Beek et al. 1991). But it is important to note that most of these schemes using log P or bioaccumulation data from fish.

It is also important to note, that patterns of bioaccumulation by terrestrial organisms cannot be predicted from behaviour of aquatic organisms (Connell 1991).

13.5 References

Beek, B., Böhling, S., Franke, C. and G. Studinger (1991). Bioakkumulation - Bewertungskonzepte und Strategien im Gesetzesvollzug, Umweltbundesamt Berlin

Butte, W. (1991). Mathematical Description of Uptake, Accumulation and Elimination of Xenobiotics in a Fish/Water System. In: Nagel, R. and R. Loskill (Eds.) Bioaccumulation in Aquatic Systems - Contributions to the Assessment, VCH Verlagsgesellschaft, Weinheim FRG, 29-42

Caspers, N. and G. Schüürmann (1991). Bioconcentration of Xenobiotics from the Chemical Industry's Point of View. In: Nagel, R. and R. Loskill (Eds.) Bioaccumulation in Aquatic Systems - Contributions to the Assessment, VCH Verlagsgesellschaft, Weinheim FRG, 81-98

Chiou, C.T. (1985).Partition coefficients of organic compounds in lipid-water systems and correlations with fish bioconcentration factors. *Environ. Sci. Technol.* 19, 57-62

Dobbs, A.J. and N. Williams (1983). Fat solubility - a property of environmental relevance? *Chemosphere* 12 (1), 97-104

Frische, R., Klöpffer, W. and W. Schönborn (1979). Bewertung von organisch-chemischen Stoffen und Produkten in Bezug auf ihr Umweltverhalten - chemische, biologische und wirtschaftliche Aspekte. Umweltbundesamt, Berlin

Halfon, E. and M.G. Reggiani (1986). On ranking chemicals for environmental hazard. *Environ. Sci, Technol.* 20 (11), 1173-1179

Hawker, D.W. (1990). Description of fish bioconcentration factors in terms of solvatochromic parameters. *Chemosphere* 20 (5), 467-477

Hawker, D.W. and D.W. Connell (1989. Factors affecting bioconcentration of trace organic contamination in waters. *Water Sci. Technol.* 21 (2), 147-150

Hryk, R. (1990). Kinetik und Metabolismus von Phenol bei den pulmonaten Schnecken *Cepea nemoralis L.* und *Lymnea stagnalis L.*. Diplomarbeit, Universität Mainz

Korte, F. (1980). Zur Frage der ökotoxikologischen Bewertung von Umweltchemikalien. GSF-Bericht Ö-509, Gesellschaft für Strahlen- und Umweltforschung, München

Kristensen, P. and H. Tyle (1991). The Assessment of Bioaccumulation. In: Nagel, R. and R. Loskill (Eds.) Bioaccumulation in Aquatic Systems - Contributions to the Assessment, VCH Verlagsgesellschaft, Weinheim FRG, 189-228

Maki, A.W. (1986). Design and application of hazard evaluation programs for the aquatic environment. In: Lloyd, W.E. (Ed.): Safety Evaluation of Drugs and Chemicals. Hemisphere Publishing Company

Mckim, J.M. and P.K. Schmieder (1991). Bioaccumulation: Does it Reflect Toxicity? In: Nagel, R. and R. Loskill (Eds.) Bioaccumulation in Aquatic Systems - Contributions to the Assessment, VCH Verlagsgesellschaft, Weinheim FRG, 161-188

Mearns, A.J. (1985). Biological implications of the management of waste materials: the importance of integrating measures of exposure, uptake and effects. In: Cardwell, R.D., Purdy, R. and R.C. Bahner (eds.): Aquatic Toxicology and Hazard Assessment: 7th Symposium. ASTM Special Technical Publication 854. Philadelphia, Pa., 335-343

Nagel, R. (1988). Fische und Umweltchemikalien - Beiträge zu einer Bewertung, Habilitationsschrift, Universität Mainz

Nendza, M. (1991). QSARs of Bioconcentration: Validity Assessment of log P_{ow}/log BCF Correlations. In: Nagel, R. and R. Loskill (Eds.) Bioaccumulation in Aquatic Systems - Contributions to the Assessment, VCH Verlagsgesellschaft, Weinheim FRG, 43-66

Opperhuizen, A. (1991). Bioconcentration and Biomagnification: is a Distinction Necessary? In: Nagel, R. and R. Loskill (Eds.) Bioaccumulation in Aquatic Systems - Contributions to the Assessment, VCH Verlagsgesellschaft, Weinheim FRG, 67-80

Opperhuizen, A., Serne, P. and J.M.D. van der Steen (1988). Thermodynamics of fish/water and octan-1-ol/water partitioning of some chlorinated benzenes. *Environ. Sci. Technol.* 22 (3), 286-292

Spacie, A. and J.L. Hamelink (1982). Alternative models for describing the bioconcentration of organics in fish. *Environ. Toxicol. Chem.* 1, 309-320

Stephan, C.E. (1985). Are the "guidelines for deriving numerical national water quality criteria for the protection of aquatic life and its uses" based on sound judgments? In: Cardwell, R.D., Purdy, R. and R.C. Bahner (eds.): Aquatic Toxicology and Hazard Assessment: 7th Symposium. ASTM Special Technical Publication 854. Philadelphia, Pa., 515-526

Wänke, D. (1991). Der Metabolismus von Phenol bei *Orconectes limosus* und *Daphnia magna*. Diplomarbeit, Universität Mainz

Index